MYSTERY ANIMALS OF THE BRITISH ISLES

Northumberland and Tyneside

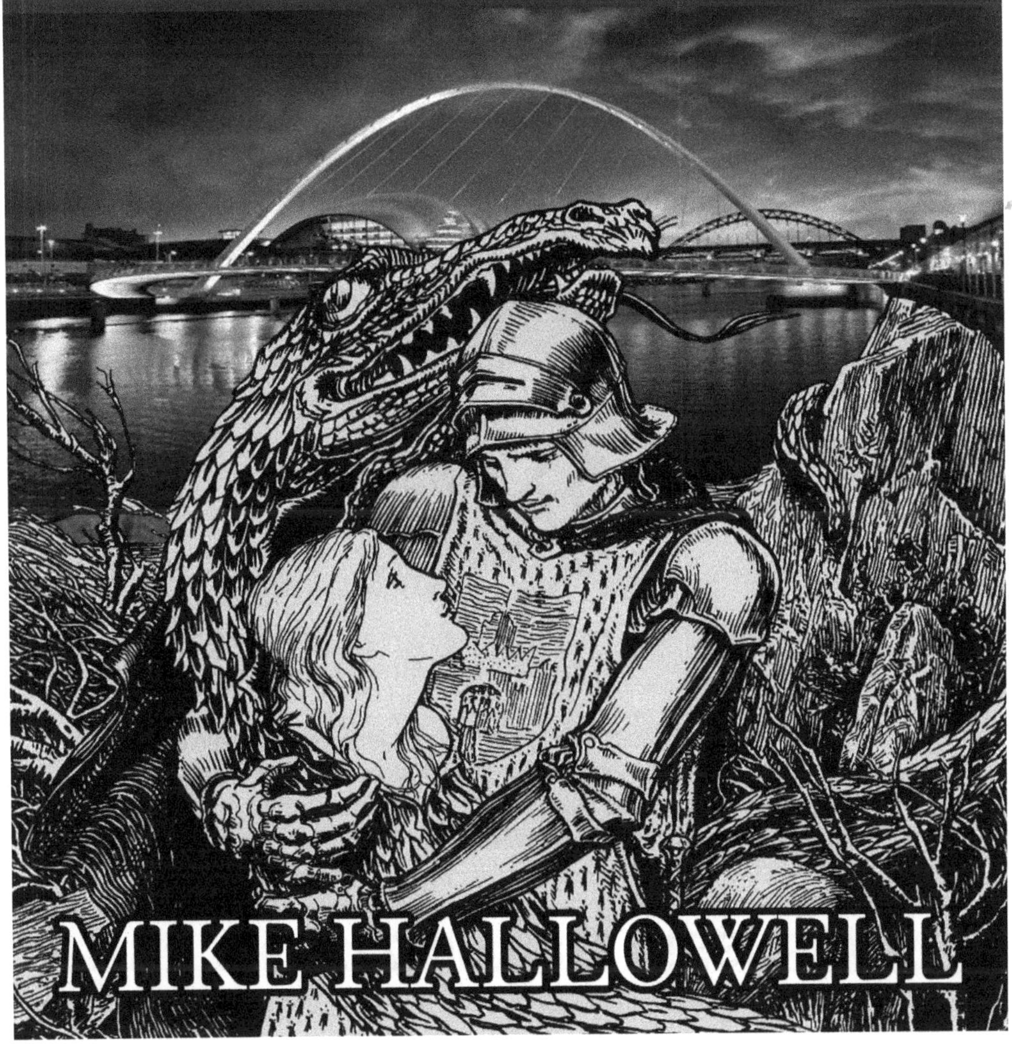

MIKE HALLOWELL

Typeset by Jonathan Downes,
Cover and Layout by *Felis citronella* for CFZ Communications
Using Microsoft Word 2000, Microsoft , Publisher 2000, Adobe Photoshop CS.

Photographs © 2008 CFZ except where noted

First published in Great Britain by CFZ Press

**CFZ Press
Myrtle Cottage
Woolsery
Bideford
North Devon
EX39 5QR**

© CFZ MMVIII

All rights reserved. Without limiting the rights under copyright reserved above, no part of this publication may be reproduced, stored in or introduced into a retrieval system, or transmitted, in any form of by any means (electronic, mechanical, photocopying, recording or otherwise), without the prior written permission of both the copyright owners and the publishers of this book.

ISBN: 978-1-905723-29-4

In Memory of David Neary
We should have talked more often.
One day we will again, my friend.

The reader should remember that some of the geographical locations mentioned in this book may not be accessible to the public. Always seek the landowner's permission before attempting to view sites of Fortean interest for yourself.

Contents

7 The Mystery Animals of the British Isles
9 Acknowledgements
11 Foreword
15 Introduction
33 Tyneside and Northumberland – Then and Now

35 The Felton Rabbit
51 The Ghost Birds of Jesmond Dene
61 Black Dogs & Howling Dogs
85 The Beast of Bolam Lake
125 The Dolly Pit Hell Hounds
131 Witch Hares & Vampire Rabbits
137 The Gast Adders of Northumberland
149 The Cleadon Panther
163 The Laidley Worm of Spindleston Hough
175 The Shony
197 Tyneside's Fee-Jee Mermaids
205 The Shrew Mouse
211 Wandering Willie
223 The Monster Lobster of Trow Rocks
231 The Dragon of Long Witton
235 The White Horse of Farding Lake

241 Conclusion
234 Bibliography

The Mystery Animals of the British Isles

More years ago than I care to remember, my first wife bought me a birthday present. It was a book about the mystery animals of Britain and Ireland, and I devoured it avidly. When I finished, I was horribly disappointed. It had covered the mystery cats of the country in some depth, as it had done with the black dog legends, and a smattering of more arcane `things` (as the late, great Ivan T. Sanderson would doubtless have dubbed them) such as the Owlman of Mawnan, and the Big Grey Man of Ben McDhui. But there was *so* much that I knew that the author had simply left out.

Where were the mystery pine martens of the westcountry? Where were the Sutherland polecats? Where was the mysterious butterfly known as Albin's Hampstead Eye? This was an Australian butterfly, the type specimen of which was caught in a cellar in Hampstead (hence the name) but no-one knows how or why? Where were the butterflies, moths, birds and even mammals known from the British Isles on the basis of a handful of specimens only? And where were the local oddities; the semi-folkloric beasts only known from a specific location.

Although at the time I had no pretensions to being a writer, I started to collect information from around the country, and with the benefit of hindsight it is probably with my disappointment with my 27th birthday present that the seeds of what would eventually grow into the Centre for Fortean Zoology were planted.

Nearly twenty years later to the day, I was sat in my garden at the Centre for Fortean Zoology [CFZ] in North Devon, sharing a bottle of wine with my wife Corinna, and my old friends Richard Freeman and Mike Hallowell. The subject of my disappointing 27th birthday present came up, and someone suggested that we do our best to redress the balance. CFZ Press, the publishing arm of the CFZ, has become the largest dedicated fortean zoological publishers in the world, and we are now in the position to put my vague daydreams of a couple of decades ago into action. We decided that rather than trying to publish one enormous tome covering the mystery animals of the whole of the British Isles (which, by the way, geographically, if not politically, includes the Republic of Ireland, but excludes the Channel Islands) we would be much happier presenting this vast array of data in a series of books, each covering a county or two. Then we realised the enormity of what we were proposing: The series would probably end up being something in the region of forty volumes in length!

However, never ones to back away from a challenge, we decided to go ahead with the project, and now - six months later - the first books in the series are being published.

It seems fitting, that - as he was there at the inception - Mike Hallowell should have the honour of being the author of the first book in the series. I am glad that he is, because it is a stonker!

We argued the toss for months over how we were going to format the series. For a long time we were intending to have a rigid format for all the books, somewhat akin to the Observer's books of the British countryside. But then we decided `No`. There are as many kinds of researcher as there are mystery animal, and it would - we felt - be more in keeping with the ethos of the CFZ, if we allowed each researcher to present his or her findings in their own inimitable style. The books, therefore, will reflect the character of the individual author.

Some will be poetic verging on mystical. Some will be matter of fact scientific. Some will be from the point of view of a naturalist, and some from the point of view of a folklorist. Some will be short, some will be long. Some will be full of scientific theorising, and some full of metaphysical speculation. But one thing is sure: Whoever gets one of these volumes for their 27th birthday present…..

…..They won't be disappointed!

Jonathan Downes
Director, Centre for Fortean Zoology
Woolfardisworthy,
North Devon,
May 2008.

Acknowledgements

Writer and researcher, Alan Tedder, without whose assistance this book could not have been completed. Darren W. Ritson, for his friendship, encouragement and kind permission to use several photographs. The staff of *The Shields Gazette*, including the editor John Szymanski, who over the years have kindly given me access to their archives. Without their assistance several stories contained in this volume would never have seen the light of day. Also thanks to the *Gazette* for permission to reproduce newspaper cuttings. *The Sunderland Echo*, for permission to reproduce newspaper cuttings. *The Western Morning News*, for permission to reproduce newspaper cuttings. Martyn and Rachel for ferrying me to the most obscure places without complaint. Fellow wordsmith, good friend and investigator Diana Jarvis for writing the foreword. Ivor Muncey and Malcolm Urquhart, for supplying me with news clippings about the dreaded Shony. Martin Embleton, Bob Brown, Corinna Thorburn, Jim Grey, Alex Carrington, Gary Smith and the many other readers of my *WraithScape* column who gave me vital leads in my hunt for the legendary Wandering Willie. Keith Bardwell of South Tyneside Central Library, for kindly helping me research aspects of the "Monster Lobster" of Trow Rocks story. And many thanks to my brother Graham for drawing the exquisite maps.

Serif, for supplying specialist software used in both the preparation of this book manuscript and the photographs contained therein. Also for their unstinting technical support when I was in a panic. (Special thanks to Robert, Ian, Jamie, Joe and Sanjay). Others, too many to mention, who contributed in many ways to the construction of this volume.

Foreword

"There are more things in heaven and earth, Horatio, than are dreamt of in your philosophy."
Hamlet *(I, v, 166-167)*

In a society that currently considers the antics of so-called celebrities as most mysterious, it is a relief to see that the world's mysteries are not always to be found in Hollywood trivia, but are no less mysterious and fascinating for that. It is even more satisfying that many of these (as yet unsolved) mysteries are to be found around the corner, whether you live in Bolton or Birmingham, Perth or Penzance. If you are lucky enough to live in the North East, then the number of sightings of unfeasible fauna increases. These mysteries are real, important and witnessed by people that are just as real and important. People like you. Unlike the celebrity media hype that fills our tabloids daily – much of it a rehash of a grain of truth - *cryptozoology* (the search for animals believed to exist, but for which conclusive evidence is missing) is the stuff of legend.

It is no surprise to me, as an honorary Northerner*, to find out that the numbers of strange and inexplicable sightings of mystery creatures in Northumberland and Tyneside are many. When the author sent me a copy of this book I found myself reading it with fascination and wonder, as no doubt you will too. However, disbelief followed quickly after. Is there really a dragon in the sea near Marsden Grotto? Is there a beast at Bolam Lake? Could

---

*I've lived above Watford Gap for at least 10 years now, so I claim the honorary title and defend my right to it, citing my (dubious) contribution to the North East's ghostly history, and therefore, its tourism, as evidence. Call me a Southern Jessie at your peril!

The mystery animals of Northumberland and Tyneside

Wallace and Gromit's were-rabbit really have been inspired by a true story? Perhaps the residents that contributed their testimonies to these accounts had had a little too much Newcastle Brown, or perhaps they were simply seeking their fifteen minutes of fame?

Of course, logic demands we try to find scientifically acceptable solutions for things we do not understand. In my time as a ghost hunter and "speaker with spirits" I have questioned my own eyes, my own ears and my own sanity on numerous occasions. In my chosen profession, I too have encountered things that can't easily be explained away and should not, in our scientific society, exist. A malevolent spirit has kept me imprisoned in a hotel room in Poole, I have heard the disembodied screams of a long dead woman in Portsmouth, and I have been caressed by the spirit of an Elizabethan merchant adventurer in Rochdale. This is the stuff of tales by the fireside that only children might dare to believe (I still have trouble believing, and I was there!) and so are the tales of the mystery animals of the North.

However, we can't just shut the book on these stories and call them fiction. We may think them fanciful and believe that the witnesses are perhaps a little sad, but their testimonies won't go away. There are simply too many testimonies, they are too well documented and fantastical creatures continue to appear, despite our best attempts to dismiss them. When these occurrences are close to home they are even more real to us, because we know these places and, in some cases, we know the witnesses. These sightings are not just things that happen to "other people" – they are things that happen to *us*.

When Mike Hallowell wrote this book, he was following in the age-old tradition of storyteller, but with a modern twist. Mike is no fantasist and he does not embellish. He is thorough, methodical and at all times he keeps to the facts, which is what makes *Mystery Animals of Northumberland and Tyneside* so very readable.

The further you progress through the stories of strange creatures on your own doorstep, the more you will realise that if there was a rational explanation to be found, Mike would have found it. He does paint colourful pictures of the areas listed and he does use interesting and graphic prose to illustrate the creatures that have been sighted. But what he does not do is stray from what has actually been documented.

I first encountered Mike when I became editor of *Vision* magazine. Mike was one of *Vision*'s most respected and talented columnists, and he made an impression on me from the day I started as editor. Within months of my first contact with him we were firm friends. I was, and I still am, continually impressed with his respect for life and death. I am also heartened to find an individual that displays such honesty, such integrity and such considerable literary skill. If Mike says he saw a thing – you know it was there. If Mike says something exists – believe me when I tell you that it *does* exist. If you are ever lucky enough to make Mike's acquaintance, you will never doubt his word and you too will know that he does not write about anything that he cannot back up. His word is his bond.

I sincerely hope that when you have worked your way through this plethora of peculiarities you will realise that there are indeed more things in heaven and earth … and also realise that you won't have to tour either to find them.

More importantly, thanks to this book, you can rest (as I will) in the knowledge that when Jade Goody's exploits have faded into obscurity, the appetite of the Felton Rabbit will still be legend, the ghostly avians will still fly over Jesmond Dene and the Howling Hounds will still haunt the area around Cragside. Perhaps one cold, wet and (naturally) stormy night, you may be telling your children that if they are sitting comfortably you will begin - and then assure yourself of a quiet night when you finish the stories in this book with "and the thing is, this *really* happened!" ...

Diana Jarvis
Editor, *Vision* Magazine (2005-2007), Ghost Hunter.

Introduction

I used to be a normal cryptozoologist - fascinated by the *yeti*, intrigued by *sasquatch* and occasionally even drawn into the seductive, aquatic lair of *Morgawr*. Then I met the infernal Mr. Downes, damn his hide, and his partner in crime Mr. Freeman. Damn Mr. Freeman's hide too, I say, but not quite as much.

Truth to tell, it was Downes to blame. He it was who took advantage of me whilst drunk and introduced me to his rather arcane ideas about zooformity. Or perhaps it may have been zooformation. Or even zooiness. I care not. The damage has been done, and is irreversible. Philosophers speak of Occam's Razor. Henceforth I shall mutter darkly about Downes' Cudgel.

To be fair, Mr. Downes' and Mr. Freeman's entrance into my career was made with impeccable timing. South Tyneside was basking in the publicity of its first cryptozoological event in many a moon, and I thought I had the whole thing sewn up as tightly as, well, an extremely tightly sewn up thing. A Turkish wrestler's jockstrap, perhaps. Not so. Mr. Downes and his notions proved to be my unravelling – and, enigmatically, my ultimate salvation. For that reason I now speak highly of Mr. Downes and no longer feature him in dark rituals carried out at midnight. The pins and powders are put away, and Downes' portrait once again hangs proudly in our hall. God bless Mr. Downes, I say, and all who sail upon him. And Mr. Freeman too.

Now, down to business.

A little over a century ago, a lady who resided in Newcastle upon Tyne was walking to a nearby steam laundry one morning when she saw a cat. The cat had four legs, a head, a tail and a relatively-proportioned torso. It walked like a cat, meowed like a cat and was, in most respects indeed the very epitome of catishness; apart from its blue fur, that is.

Cats do not normally have blue fur. A tinge or tint of blue, or even a slight hue of blue, perhaps. But this cat was *bright* blue; the sort of blue that one may expect to see on a dress worn by Dame Edna or emanating from a bank of 1970s disco lights in a seedy nightclub. Cats of neon blue are not the norm, especially when they disappear in a flash of light as this one did. The mystery was never solved.

Mystery animals are big business, or at least would be if we could capture them on a regular basis. The first person that secures a *yeti*, as it trundles across the Himalayas, and displays it in a cage in Times Square, will probably make a pretty penny. Any hunter who kidnaps a *sasquatch*, and delivers it to the offices of the *National Enquirer* will probably be able to retire soon thereafter, troubled by nothing more than his, or her, next book deal. Even the Loch Ness monster, which is in danger of becoming passé within the world of strange beasts, will earn its captor a small fortune, and the Freedom of City of Edinburgh. Or maybe even Govan. If it's ever captured, that is.

The world is a pretty big place. True, its not as big as, say, Jupiter, but it's big enough to hide in if you're on the run. The Inland Revenue will always be able to find you, of course, but they'd find you even if you *were* on Jupiter, so it makes no difference.

Aeons ago on planet earth, there was a time when there were no humans. All manner of strange fauna tramped the globe, but humanity had not yet made its dubious entrance on to the world's stage. Some would argue that the earth would have been better if we hadn't made an entrance at all, but that's another issue.

But then we humans did arrive – somehow – and the number of species we shared the globe with began to shrink. Of course, many of them disappeared without our help. Others, like the carrier pigeon, became extinct with our unwelcome assistance. Despite the fact that we now exercise almost complete dominance over our planet, there are still a few uncharted regions where previously unknown species may exist. It's a dead certainty, in fact, as hardly a week goes by without a new critter being found in some far-flung clime. Its also likely – highly likely – that many species we currently think to be extinct may not be. The Tasmanian thylacine is but one example.

Of course, the creatures hitherto discussed all belong firmly in the physical realm, and these are the ones that almost everyone thinks of when we talk about "animal life". But what of other, less corporeal beasts that reportedly invade our world from time-to-time before supposedly teleporting back to whatever dimension they hailed from? Oh, such critters do exist, believe me. Some are always ethereal and never seem to enjoy the solidity of more conventional species. Others may appear solid and look for all the world like an ordinary animal until they do something quite extraordinary, like floating through the air or disappearing in a puff of luminous smoke. Such creatures are never given a place within the conventional classifications or taxonomies of recognised fauna, for orthodox science simply refuses to acknowledge they exist at all. But they do, and you will read about many of them within the pages of this book.

Some years ago I investigated a strange case, which involved an ape-like creature, which was dubbed the Beast of Bolam Lake. I'll inform the reader of this investigation presently in considerable detail, but at this juncture it may be appropriate to point out one or two conundrums associated with the appearance of this strange creature. Firstly, it was seen in an area of woodland that was simply too small to house a breeding population of large mammals. Secondly, the idea that such a population could exist in such a small area without being found and captured seemed impossible. These facts in themselves indicated that the Beast of Bolam Lake, whatever it was, was not a conventional animal.

Centuries ago, the northeast of England was largely covered by huge woods and forests. Back then, the idea that unknown species could have existed there for untold ages without discovery was not unlikely. Now, however, the forests and woods of the region have been decimated by the onslaught of urban life. Huge cities and towns now occupy areas that at one time were filled with nought but trees, shrubs and moors. Although the possibility still exists that the small areas of forest and woodland in the north could indeed house unknown species of animals other than very small ones, it is a remote one. And yet, reports still come in of strange beasts that to all intents and purposes simply shouldn't be there.

Until the turn of the 21st Century the north east of England was really uncharted territory when it came to paranormal research. A handful of 19th and early 20th Century writers had written books about the mysteries of the region, but little of what they had to say ever reached a broader audience. To try and rectify matters I launched *The Twilight Worlds Paranormal Research Society* in South Tyneside. The organisation's brief was a broad one; to investigate all manner of strange occurrences in the northeastern area. These occurrences included UFO sightings, hauntings, psychic phenomena and, of course cryptozoological sightings. Less than two years earlier, I had been asked to write a regular column for *The Shields Gazette*, the country's oldest provincial newspaper. The column, *Bizarre* (now called *WraithScape*), focused on much the same sort of stuff, but I quickly became aware that there were literally hundreds of strange accounts that were in danger of being overswept by the sands of time. Hence I regularly revived such stories within my column for a whole new generation of readers who were unaware of them.

I've had a number of cryptozoological encounters myself. Not all of them have taken place within the geographical scope of this volume, but I have decided to include two for the benefit of those who are taking their first dip into the turbulent waters of this arcane field of endeavour.

In the autumn of 2003, I spent some time living on tribal land in Webster Parish, Louisiana. Bigfoot, or s*asquatch*, has been seen many times in Louisiana and this southern State is actually one of its favourite stomping grounds.

One day I went for a walk through the woods, and at some point decided to return to the lodge where I was staying for something to eat. I walked back to the lodge, but then changed my mind and decided to walk some more. I carried on up another path, which took me past a rather picturesque wood cabin with an old-fashioned stove chimney pointing skyward. I passed one of the tribe's tractors – used to level off the land and for a multitude of other things – and found myself in a small clearing. I paused. Something was different here. I heard a cracking noise to my right, emanating from what appeared to be a large expanse of dense woodland and bush foliage. In fact, it was so dense that it was impossible to see more than a few feet into the interior. I concluded that the noise must have been made by a relatively large creature, and I confess to feeling apprehensive at that juncture. Nevertheless, I carried on walking and noticed another, smaller clearing also to my right. To my amazement, I saw something that I had only seen twice before; once in a book and once in real life. There, before me, was what I clearly recognised as a *sasquatch tipi*.

Sasquatch – often referred to incorrectly as a yeti - is a large, bipedal hominid of unknown determination. For centuries this mysterious creature has stalked various areas of North America. It has been seen thousands of times, chased hundreds of times but never, to my knowledge, caught. For those who are unaware of the legendary sasquatch, in general terms it is between six and eight feet tall, extremely muscular and covered (except for the facial area) from head to toe with thick black (or sometimes chestnut coloured) hair. Those privileged to have seen this mysterious man of the woods are usually most stricken with how human he looks apart from his hirsute appearance.

So many sasquatch sightings have been recorded that it is, in my opinion, bordering on the delusional to deny the creature's existence. Sasquatch usually stays away from humans, but has been known to become aggressive if either startled or challenged.

Like all creatures that have not been subject to scientific scrutiny, the behaviour patterns of sasquatch are poorly understood. One of his strangest habits is to indulge in the creation of "sasquatch tipis". These are dome-like affairs created by bringing together the tops of saplings and either tying them together or interlacing them. Indeed, they really do look like the framework of an American Indian dwelling without the covering, although not, strictly speaking, a tipi. No one knows why sasquatch does this, or whether the "tipis" have a practical, cultural or spiritual function.

I have seen such "tipis" before, but, ironically, not in the USA. I first saw them in rural Northumberland, England, not too far from my home. As a paranormal investigator, several colleagues and I were, at the beginning of 2003, asked to investigate sightings of a sasquatch-like creature near an area known as Bolam Lake. We concluded that there certainly was a hairy hominid of some sort stomping around the woods, but unanimously agreed that it just couldn't be a flesh-and-blood creature in the normal sense of the word. The area was too small for a breeding population – or even a single animal – to hide in indefinitely. Further, as a creature of that size would need a calorific intake of around 9,000kc per day, there just wasn't enough food around to sustain it. We concluded, then, that the beast was "real", but probably of a psychic or spiritual nature and could both appear and disappear at will.

Such otherworldly creatures are, in cryptozoological terminology, known as "zooforms", and one of their features is that they may, when present, leave physical traces as real as any conventional beast. The Beast of Bolam, as it became known, certainly left traces. We photographed a number of startling (and very large) footprints. We also found several classical "sasquatch tipis". The beast, then, shared a bizarre but positively identifiable link with its North American cousin. That is why, when I saw the "sasquatch tipi" in the woods on the land, I knew exactly what I was looking at. The question was, when had sasquatch last been seen around here?

Although feeling distinctly nervous, I walked into the clearing towards the tipi. I felt as if I was being watched. This tipi was old, not fresh. One sapling was bent over so severely it was virtually touching the ground. As far as I could tell, it was not being held in place. It looked as if it had either grown that way – which would have been very odd – or had stayed that way after the other aspects of the tipi had collapsed or fallen away. I examined the tip of the small tree and found the remains of other branches – long dead – intertwined with the living foliage.

I left the clearing, and as I did so I heard another sharp crack, as if something heavy had stood on – and snapped – a thick branch. The feeling of being watched intensified. I quickly took some photographs and then departed.

I walked back to the lodge, and found my friend sitting at the computer checking his e-mail.

"You ever seen anything...well, funny here?"

"Like what...spirits".

"Like Sasquatch".

"Let's take a walk, brother", he said.

We strolled up past the picturesque log cabin and through the clearing. We passed the smaller clearing on the right, which contained the sasquatch tipi. I wondered if my friend had seen it, and if so, recognised it for what it was. If sasquatch had made this, then he may not have wanted it to be found or seen. Out of respect for my sasquatch brother, I felt that I should not draw my friend's attention to his work. I did not, therefore, mention it to him.

We walked on, and the ground fell away to the right. My friend asked me to walk in his footsteps. There were potholes, soft earth and snakes, any one of which could prove to be problematical. We were now, I would estimate, no more than one hundred yards away from the sasquatch tipi behind us. Below us ran a beautiful creek.

"I saw him here once".

"Who?"

"The yeti".

"Sasquatch?"

"Sasquatch...yeti...whatever you wanna call him. I saw him, just there..." he said, pointing. *"One other time I saw a whole bunch of them walkin' across a field over there. Some of 'em were young. They were a family, I think, maybe 7 or 8 of them".*

This was too much to be a coincidence. My friend had seen sasquatch very close to the tipi – a sasquatch tipi which, as far as I could tell, he was still unaware existed. Both his sighting and the presence of the tipi were acting as witnesses for each other. I knew then that he was telling the truth; not that I ever doubted it. Every morning, after that, I walked to that clearing at dawn, hoping that my sasquatch brother would show up. I never saw him, but on no less than three occasions I heard something moving about, crashing around. I also felt, yet again, as if something was watching me intensely.

The day after I arrived, shortly after my colleague had shown me where he'd encountered sasquatch, he decided to take me to a bayou. He also took his pride and joy – a beautiful boat that he'd built himself with help, I think, from one of his daughters. We loaded up the boat on the roof rack. He showed me how to tie a number of different knots with the blue twine, which he used to secure it. Then we got in the car. Little Wolf – my friend's *dawg* - sat in the back, we having decided to take him along for a treat.

We were heading for Dixie Inn and the Dorcheat Bayou, but I didn't notice the scenery too much. My mind was totally focussed on the conversation I was having with my colleague.

"See, it's like this, brother. Tribe is family. That's the trouble with the world today. Everyone is broken up. Everyone is consumed with the satisfaction of self instead of the service of others. It's so sad."

And my friend was right, of course. Self-sacrifice, absolute trustworthiness and humility, along with a multitude of other good qualities, are considered "old hat" now. Although I didn't know it, I'd find out more about these qualities within the coming weeks than many people do in a lifetime.

We arrived at the Dorcheat Bayou and parked the car near a small cluster of shops and restaurants. Across the road was a small jetty, which intruded into what was, quite simply, the most amazing scene of absolute tranquillity. Huge trees of every hue and colour towered over a river, which was so still its surface was a perfect mirror. I had never, ever seen anything so beautiful before.

We untied the boat and carried it down a cement incline to the water's edge. My friend then secured it to the jetty. There was a young, heavily-built man wearing a baseball cap sitting on the jetty. At his feet sat a strange-looking dog. My friend engaged him in conversation, and I joined in moments later.

"What sort of dog is that?" I enquired. *"I've never seen anything like it."*

"Well", said the man, *"It's an Australian heeler. They do look pretty weird, I guess, but they're good dogs."*

We chatted for a while, and then my friend pointed out that there wasn't much daylight left.

Not having knee-length waterproof boots, I walked along the jetty and climbed down into the canoe. I don't swim and have never had the remotest inclination to be a mariner. I am, in fact, a dyed-in-the-wool landlubber. The reader will understand, then, that I was distinctly nervous.

A number of questions entered my mind. Not much more than a day earlier, I had been sitting at home with a nice cup of tea in the comfort of my own living room, at peace with the world. Now, just a heartbeat later in cosmic terms, I was half way around the world in a strange country that, at least in this state, enjoyed a sub-tropical climate. More than that, I was about to embark upon a journey down a beautiful but eerie river with an Indian whom I really liked but hardly knew. What on earth was I doing here? Was this, I wondered, a crazy dream from which I would soon wake up? Would I, in a moment, find myself lying next to my wife Jackie in bed, waiting for the alarm to go off, heralding another day in my office?

My colleague waded through the water from the ramp and entered the boat, handing Little Wolf to me as he did so. He also handed me a paddle. He then untied the canoe and pushed us away. I paddled to the right, my colleague to the left. I made a mess of things at first, but soon got the hang of it. And so we departed – an Indian with a head full of tribal lore and a Geordie lad from Tyneside with a hyperactive wolf cub between his legs.

We turned left, and paddled towards a large bridge, which straddled the bayou. Its concrete legs rose from the water like giant arrow shafts. My colleague let me paddle on my own for a while, until it became obvious that we were heading directly for one of the aforementioned pillars. No matter what I did with my paddle, the boat steadfastly maintained its collision course. My friend started to paddle too, and deftly steered us around it.

The water was almost black, and its surface glistened like molten tar. In the distance I heard a bird cry, "Caw! Caw!" and for some reason it made me feel nervous.

Native Americans believe that rivers are actually sentient life forms, and often refer to them as the Long People. This will sound strange to Europeans, and I understand why. In the UK, most rivers that you would canoe on are bubbling, flowing and clear to a degree. There are living things *in* the water, most Britons would agree, but few would suggest that the water itself possessed consciousness. It is a vehicle that *carries* life, but is not life itself, they would say.

The bayous *are* different. The water is shadowy, mysterious, impenetrable, and it guards its secrets well. The just-visible greenish luminescence guilds the dark stillness of the bayou, making it appear almost solid. That illusion of solidity continues where bayou meets bank, the water and the earth blending with each other seamlessly in the shadows. The hush of the bayou melds with the silence of the woods, the colours of the bayou with the tint of the trees. Everything here is alive, and exudes the life essence audaciously, not subtly as in England. There are myriad life forms, animal and vegetable, within the water and on the banks, but the entire bayou possesses a collective consciousness, which is a synergy of all the others that dwell there.

As the canoe sailed silently ahead, I firmly came to believe that the bayou itself, and in all its aspects, was truly alive. I also felt that it had a tremendous power, and that it could, with one small snap of its metaphorical fingers, eradicate me if I didn't show it proper respect. Eventually we came to a fork. The bayou widened measurably, but was cut into two separate channels by an island that lay dead ahead.

"Left or right?"

"Sorry?"

"Which way d'you think we should go? Let your instinct tell you."

"Let's go right", I responded. The trees seemed to be getting even larger, and I had this strange feeling of being cocooned in a blanket of living green.

"I want you to stop paddling", said my colleague. *"Just look around you. Be observant. You need to use your eyes more."*

The canoe glided effortlessly, the edge of the woods on my right, the island on the left.

"See on that island? Used to be a man lived there. Least, he lived there till he died."

I was amazed. How could anyone *choose* to live on a remote island in the middle of this eerie place; a place where, I was told by one good ol' boy, "nature is just waiting for a chance to bite you in the ass"?

"He must have had some guts living out here on his own", I said.

"Maybe... maybe..." replied my friend dreamily.

I thought he was thinking about something, but he wasn't. I turned and looked at him. His eyes were staring firmly ahead, and he raised a hand to indicate that silence was necessary. Then he pointed.

"Alligators – over there".

Fascinating, I thought. As long as they *stay* over there...

Although the bayous are intimidating, they are also, paradoxically, places of great tranquillity. On the land, particularly of an evening, the woods came alive with noisy insects. On the bayou, at least before sundown, there is a silence that is so intense you can almost touch it. That silence is occasionally broken by a screech, a caw or a howl, but there is no background noise as such.

My friend started to talk again.

"Learn to use your eyes, brother. You need to use your eyes more. Look around you. What do you see?"

"I see great beauty. I see life".

"Good. But what do you really see?"

Momentarily I was puzzled. What else *was* there to see? But then I looked around me again. This time I allowed spirit to guide my eyes, to help me see things anew. Then I knew exactly what my friend was trying to get me to do. My "looking" was a European way of looking. My brain was simply digesting shapes, dimensions and colours. He was trying to get me to see not only things, but also the things *behind* those things. He wanted me to see how things changed, how things differed. He wanted me to see minute movements, not just obvious ones. When I looked again, I saw the bayou through different eyes.

I could detect movements much more easily. The water was not really still at all. It was just that the movements on its surface were delicate, subtle, easy to miss. I could see ripples so faint they were almost invisible. Had something extremely small – an insect, perhaps - entered the water, or was something underneath almost, but not quite, breaking the surface?

My contemplative meditation was rudely interrupted by Little Wolf. Without warning, this puppy – who has a true warrior spirit – leaped over the side into the bayou. Stunned, I watched as he swam happily around in the deep, dark, waters. I glanced at my friend, and could see in an instant that he was alarmed. Quietly but firmly he spoke.

"Now we really have a problem. We have to get that dog back in the boat, and we have to do it now."

"What's wrong?"

"Alligators. Alligators just love dog. A dawg is their favourite snack."

I asked my colleague if he wanted me to paddle, but he said no. Expertly he manoeuvred the canoe until Little Wolf came within arm's length. Then, he quickly leaned over and, quite literally, pulled the puppy out by the scruff of its neck.

My friend paddled the canoe to the bank. There were huge trees, lined up like soldiers, and their branches hung over the bayou like protective arms. Dangling from them were huge drapes of moss – a soft, blue-grey material that was pleasant to the touch.

"Quickly", urged my colleague, "*Grab as much of that stuff as you can, then make a nest of it in the bottom of the boat. Quickly, now".*

I did as he requested, although standing up in the canoe was scary. I found it difficult to balance, and fully expected to end up in the water. As soon as the "nest" was completed, My friend handed me Little Wolf and said, "*Wrap him up in the moss. That way he won't get cold, 'cause cold dogs whine. We don't want the dog whinin'.*"

I had an awful feeling that I knew what my pal was getting at. Was he suggesting that a whining dog would be virtually the same as a dinner bell, and that if Little Wolf whimpered we'd soon have a family of hungry alligators around expecting lunch? I didn't ask. Ignorance truly is bliss, sometimes. I was later told by someone that the moss would also mask the smell of the dog, thereby making it difficult for the 'gators to detect his presence. Whether this was true or not I really don't know, but being painfully aware that these creatures were around I was prepared to try anything that would minimise the risk of my colleague, Little Wolf or my good self becoming a 'gator snack.

I made a hole in the middle of this improvised nest and placed Little Wolf in it. Working speedily I wrapped the moss around him until only his head was visible. He whimpered faintly – once, maybe twice – and then promptly fell asleep. As we paddled away, I looked at Little Wolf's face. He was a picture of contentedness.

My friend took the incident in his stride. You could tell he was used to the bayous. I wasn't, and found the close presence of alligators – even though I couldn't see them – unnerving.

"*Light's about done. We need to turn back now.*"

We turned the canoe around and headed back the way we'd came.

"*You know, this is God's country, brother*", said my colleague, and I agreed with him.

As we neared a left-hand wheel in the bayou, he pointed out a beaver dam just ahead in what looked like a small tributary. I switched on my digital camera and started filming.

"*See there? That's where some 'gators are holed up…just waitin'.*"

I asked my pal how big the 'gators could get.

"*Oh, sometimes ten to twelve feet. It isn't common, but it's not unknown.*"

Later, someone told me that an eighteen-footer had been seen in that bayou, once.

As we retraced our passage down the bayou, I continued to look around me with what I later came to term my spirit eyes – a type of seeing which allows you to view the physical world around you in more depth, and also see a little more of the non-physical world too. To see – I mean *truly* see, as opposed to just looking – was one of the greatest lessons I learned whilst in the USA. Had I not learned this lesson quickly, I doubt I would have seen the brief movement on the tree just ahead.

The tree was not the biggest one I'd seen during our expedition up the bayou, but, by British standards, it was big enough. As the darkness was just starting to cut in I could not see in any great detail, and everything on the bank was now taking on a silhouette-like appearance. However, I distinctly saw a movement on the right-hand side of the trunk about six feet from the ground. Something shivered – crawled, slithered, I'm not sure – about two feet earthwards. Then it stopped. To me, it looked like a snake, although I couldn't be sure.

"Is that a snake?"

"Where?"

"There...over there, on that tree..."

My friend looked over to where I was pointing, and as he did so there was another movement. The creature moved earthwards again, and both my pal and I saw it. We both recognised it for what it was; a squirrel. Its feet connected with Mother Earth, and then, in an instant, it was gone; disappearing into the dense undergrowth which hadn't changed in millennia. It hadn't been a snake, then, as I'd first imagined, but I was pleased that I was now able to see things I normally would have missed.

"See?" said my colleague, *"Now you're payin' attention to things!"*

Another five minutes or so passed, and I decided to film some more of the breathtaking scenery. I depressed the "record" button on my camera. Suddenly, the setting sun squeezed through a gap in the trees, gifting us with a beautiful display of reds, golds and almost-whites.

"Hey! You see that? Look!...Look!...Just ahead! You see that?"

At first I couldn't see anything. My pal was pointing to a spot on the bank, just ahead to our right, but all I could see were trees. Then my heart jumped. Partially hidden behind a small tree trunk was something large and black.

What follows is a transcript of the recorded conversation between my friend and I at this point.

HE: *I can't see it. Unless it's gone...Oh, its there...I wonder what that is. See?...Between the trees?....The black thing?*
ME: *Oh, yeah...yeah.*
HE: *It's... it's not a bear. It would already be gone...You can only see...*
ME: *It's standing real still...*
HE: *If that were a bear...you would not...it would be gone. You only see a bear for a few seconds...*
ME: *What...What do you think that is?*
HE: *Uh...maybe it's the stump of a tree which blowed over, leaving a piece of it...?"*
ME: *But...it's just with it being darker than the rest...*
HE: *Yeah, it is pretty unique...that's why...keep your eyes peeled for anything unusual...any*

kind of hump or bump that don't fit the mainstream...
[At this juncture we heard a loud bang coming from the woods behind us.]
ME: *What was that?*
HE: *Shotgun...squirrel hunters...this is the squirrel season, and they're shooting squirrels everywhere.*

[At the moment the shotgun was fired, the "black thing" that my friend and I had both seen disappeared in an instant.]

ME: *I can't see that thing now!*
HE: *There it is!...Its difficult on this side...its [unintelligible].*

Now it was back in view. At this point, as I stared, I slipped into an altered state of consciousness. I picked up both emotions and sensations. Anxiousness, heightened awareness, an increased heartbeat and the withholding of breath to prevent movement. Within seconds I was back in the "real" world and staring at "it".

More emotions flooded my senses. We had to get away from this place. Like the sensation I got in the woods when I saw the Sasquatch tipi, I felt here too that whatever I was looking at should be left alone. I didn't believe it was a tree trunk. When that shotgun went off, it had *moved*.

The problem was the trees. As we slowly drifted downstream, our angle of vision changed continually and a succession of tree trunks passed in front of "it". At one point it seemed as if I was looking at the creature - if such it was - through prison bars. At no point could we get a clear and unhindered view of the beast.

The "black thing" disappeared again, and I verbalised my feelings:

ME: *"Pity..."*

After a moment or so I asked my friend whether it could have been bigfoot.

ME: *Maybe it was sasquatch?*
HE: *Huh?*
ME: *Could it have been sasquatch...bigfoot?*
HE: *Uh...it's [unintelligible].*

I stopped filming. I still wasn't convinced it was a tree stump at all. Whatever the thing was, I estimated that it had been between six and eight feet tall. It was – or had seemed to be - covered in jet-black hair, and was pretty much human in shape. What struck me most forcefully was how still it had been standing, as if it didn't want to be seen. There was, I concluded later, something unnatural about that stillness. I have never seen anything so still in all my life. This was a stillness that was greater than the stillness of a tree or a rock. This lack of both motion and movement went *beyond* stillness. This was stillness *plus*. I know this doesn't make much sense scientifically, but it's the only way I can describe it.

"It" was now behind us. As the canoe pulled away and the sun continued to set, the still, black shape became less distinct. Did my pal really believe it was a tree stump? Or was he testing me, getting me to use my spirit-eyes and reach my own conclusions? All I know is that I genuinely believe that what we saw was a living creature – a creature that, according to my friend, could not have been a bear. According to me it couldn't have been a tree trunk. What had we seen? I do not know, cannot say, at least with any certainty. What I do know is that I interacted with *something* standing in the woods, and not only picked up emotions and sensations but also lived through a short but intense spiritual experience. This had been too much of a coincidence, I thought. First my friend tells me how to see properly – with my spirit, and not just my eyes – after which I have my newfound ability tested by a squirrel. Then what? Just after I've learned how to see in this way, as if by magic, I have an experience that tests that ability to the uttermost. No, this was no coincidence.

Once again I voiced my feelings.

"This was meant to be a learning experience for me, wasn't it?"

"Yup. Sure was. You like it?"

"I'm not sure, but I feel as if I've been taken to another level somehow."

"Good...that's good".

It is strange indeed that I should have had this experience, and stranger still that I should have had a similar encounter across the Atlantic in England months earlier. You'll read more about my encounter with the Beast of Bolam Lake later in this book.

The other cryptozoological experience I'd like to relate to you took place not in the humid climes of a Louisianan bayou, but right next to my own home in West Boldon, England.

It was in the month of August, 1993. My youngest son, Aaron, was eight years old at the time and had been playing football with a young friend in a play area not far from our house. I was busy decorating the upstairs hallway at the time and at some point happened to glance out of the upper landing window. It was a bright, sunny afternoon and the shrubbery was in full bloom. Suddenly and without warning a huge bird flew into my line of vision. It had obviously flown over the rooftops of some houses in my street to the south, and, almost without hesitation, landed in the shrubbery directly opposite the window I was looking out of. Seeing a bird land in some bushes is not unusual, but a bird of this girth should not have been winging its way through the skies of West Boldon. I'm not an ornithologist, but the sheer size and demeanour of this creature told me without doubt that it was not a species native to the British Isles.

Let me describe it for you. It stood about three feet in height and was covered in feathers that held the colour of milky coffee. It had long legs, the feet of a wading bird and an extremely long, pointed beak. As the dense foliage completely enveloped the bird I could no longer see it, and wondered momentarily whether I could have imagined it. Just then, I looked up the street and could see my son and his friend walking down towards our small front garden. They were idly kicking a football back and forth between them. They paused only feet away from the shrubbery that was hiding the huge bird, and looked towards the house. Aaron saw me looking out of the upstairs window and waved. His friend followed suit.

I tried to open the landing window to tell him about the colossal avian hiding in the bushes behind them, but the key for the window lock was missing. I'd taken it out whilst varnishing the window surrounds, and couldn't remember where I'd put it. I decided to shout as loud as I could.

"Aaron! There's a huge bird in the bushes behind you! Look!"

"What, Dad? I can't hear you".

"THERE'S A HUGE BIRD IN THE BUSHES BEHIND YOU!"

"You're just being silly, Dad!"

"NO, HONESTLY, I'M NOT! THERE'S A BIRD IN THE BUSHES BEHIND YOU AND ITS HUGE – A GIANT!"

Aaron and his friend laughed. Then the foliage rustled and the colossus stepped forward onto a small patch of grass.

Young kids can run pretty damned fast – crowded wedding receptions are their speciality venue – but I've never seen kids run as fast as those two. They hurtled across the drive, down the path and into the house shrieking like banshees. Despite my intense puzzlement over the bird, their comical reaction made me break out into hysterical laughter. The avian interloper flapped its huge wings and took off like an airliner. I watched as it soared off in the direction from whence it had come.

A few miles from our home there is a bird wildlife sanctuary, and my assumption was that one of its feathered residents must have decided to take a short trip away for the day. I rang them and described the bird that I'd seen. I was intrigued, and wanted to know what it was. They were puzzled too, and said it didn't fit the description of any birds currently residing there. To this day the mystery remains unsolved.

One of the beauties of cryptozoological or unknown animals is that they can turn up any place, any time. The Deep South in the USA, an urban village in the north east of England…it makes no difference. This volume specifically covers the weird and wonderful animals that are said to roam Northumberland and Tyneside – the first book of its kind to do so, to my knowledge.

Eventually I parted ways with *Twilight Worlds*, as the organisation was going in a direction that I didn't much care for. I still write my *WraithScape* column, however, which is now more popular than ever. Over the years I have accumulated dozens of stories from the Northumberland and Tyneside areas relating to strange creatures which – whatever their nature – simply shouldn't be there. The Centre for Fortean Zoology, of which I am the Tyneside Representative, commissioned me to write up these accounts in book form. You now hold in your hands the result, and I sincerely hope you enjoy it.

Mike Hallowell
West Boldon, 2007

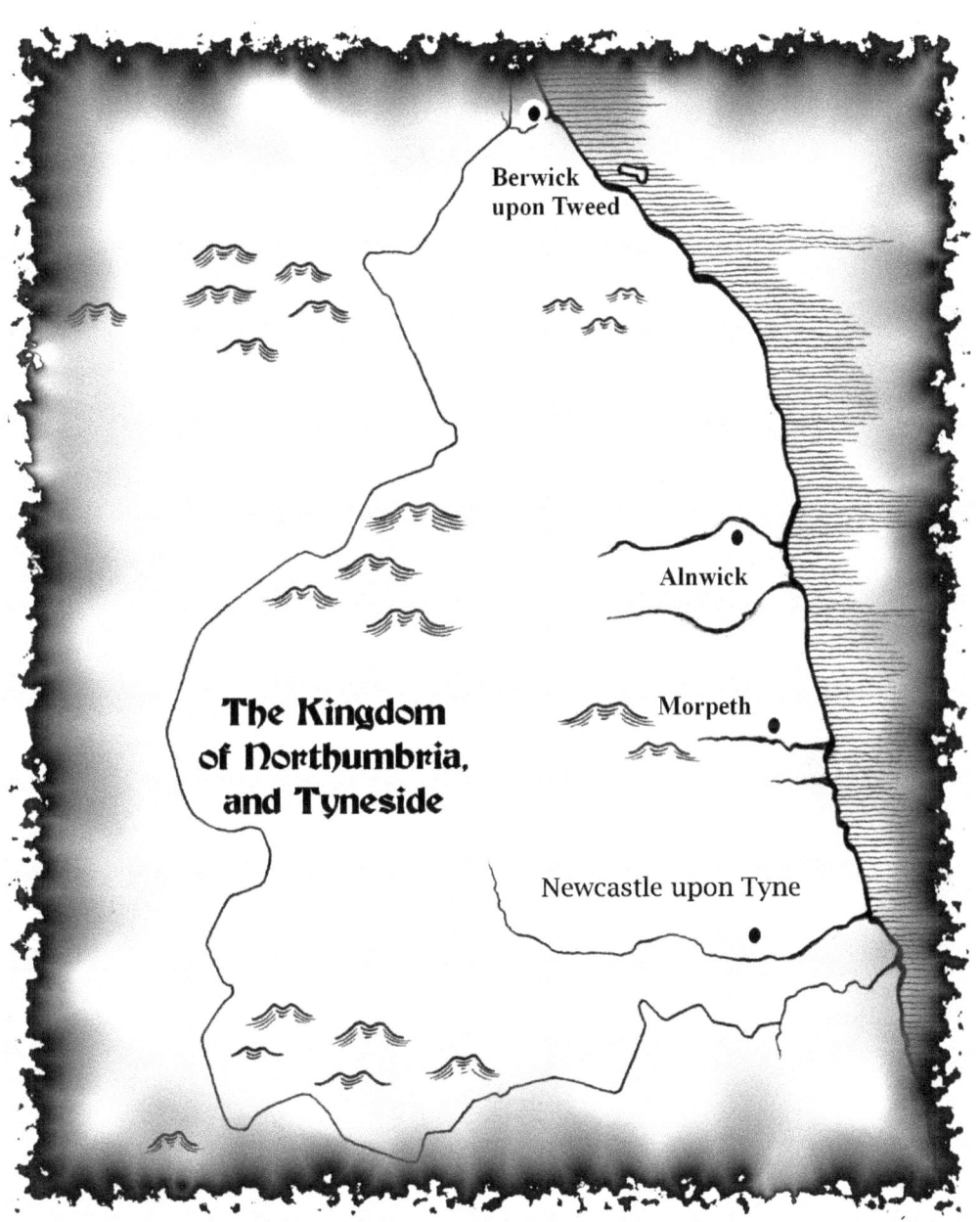

Tyneside and Northumberland – Then and Now

The first thing that strikes visitors to Northumberland about the landscape is the bumpy bits. Some places have lots of flat bits and only one or two very big bumpy bits called mountains. Northumberland has hardly any big bumpy bits, but it makes up for it by having lots and lots of smaller ones. Geordies and Northumbrians are used to all this bumpiness, and therefore feel very uneasy when journeying through a landscape with hardly any. When I visit my sons in Peterborough, I feel incredibly disorientated by all the flatness. Providing that there aren't any houses or factory units in the way, I can see for miles unhindered. In the north, it is impossible to gaze for more than three feet without being interrupted by a hillock.

Northumberland has the most underrated countryside in Great Britain. The undulating landscape adds a dimension to the rich, verdant foliage and keeps travellers in a constant state of expectation; you never know what lies around the next bumpy bit, just out of sight.

There was a time when virtually all our region was covered in forest. Some of it disappeared in part by the action of nature, such as the Hayning Wood which used to stretch from the coast at South Shields all the way to Gateshead. Changes in the climate and other factors drove the tree-line further inland, although it must be said that human intervention was by far the biggest culprit. The industrialisation of Tyneside saw the countryside devastated as huge swathes of concrete replaced woodland and pasture. In comparison, Northumberland remained relatively unscathed. The twin curses of industrialisation and urbanisation managed to gain a foothold in places like Ashington and Blyth, but overall the greenery remained in the ascendant.

The main victim of the Industrial Revolution in Northumberland and Tyneside was the wilderness. Time was when hermits could quite literally leave civilisation behind and head off into the Northumbrian wilderness. If they wanted, they could, with a little thought, spend the rest of their natural lives there without ever clapping eyes on another human being. One cannot do that now, of course. You really have to look hard to find anywhere in the region that could truly be classed as a wilderness in the proper sense of the word.

The first shock to the Northumbrian and Tyneside ecosystem came in the form of the Roman Occupation. They "civilised" us by teaching us how to farm more efficiently and introduced us to such diverse wonders as fish sauce, central heating and lettuce. All very well, but they had a mania for building walls to keep people (or maybe cryptids) in or out of this place or that. The resulting edifices were incredibly majestic, but they proved to be the first blot of urban life on a hitherto natural landscape. The rest, as they say, is history. With the walls came the roads, and with the roads the traffic. A good number of original Roman roads are still used by Northumbrians today, believe it or not.

By the time of the Danelaw, there was still a great deal of our countryside left, although the wilderness had been diminished. The Vikings gifted Tynesiders with their inherent obsession for shipbuilding, and, over the course of the next millennium, the river became one of the greatest hives of maritime construction the world had ever seen. The urbanisation of Tyneside was far greater than that which took place in Northumberland, and this change actually affected the perception of mystery animals. In the industrialised area of Tyneside, cryptids tended to be mainly zooform; strange beasts, weird in shape and of uncertain provenance. In Northumberland, "elementals" still ruled the roost, and mystery animals were essentially "normal" in appearance apart from their size – or their ability to talk and/or sing.

Over the last two decades there has been a concerted effort to re-green many areas – particularly those badly affected by industrialisation. I was raised in Jarrow, and as a child I recall it being an absolutely horrible place to live. During the last war, an American army officer allegedly called it the "asshole of the universe" - or so I've been told – and in parts it pretty much was. Now? You wouldn't recognise the place. South Tyneside is a fantastic place to live, and much of the wasteland has been reclaimed and donated back to Mother Nature. Cloth caps, back-to-back tenements, smog and whippets are out – and trees, country walks, nature trails and fish ponds are very much in.

Perhaps the greatest effort to re-green the area is the establishment of the Great North Forest – a project designed to fill up grey areas with trees. Now I'm all for creating forests, but I must confess to having a worry about this sort of exercise. Re-greening projects are often branded as "experiences" and promoters seem to need to make them "entertaining for a new generation". Let me nail my colours to the mast; I do not want a "forest experience". I do not want to be entertained by a forest. I have no interest in visiting a forest that mixes real trees with virtual ones. If we are going to create new forests, then let's do just that. I want my forest to be filled with a good variety of indigenous flora and fauna. I do not want my forest to be a theme park with a tree stuck in the middle, surrounded by visitors' centres, fast-food emporiums and playgrounds. Real forests contain magic and myth; they are places where we can go when we want to see cryptids. Up to now, the Great North Forest project seems to be doing a pretty good job – but I'm keeping my eye on them.

As respect for the countryside increases, the myth and the magic are getting stronger simultaneously. As our sense of wonder returns, so, I am sure, will those strange creatures whose essence is always just beyond our understanding. I await their arrival with eager anticipation.

The Felton Rabbit

Rabbits are mammals that belong to the family *Leporidae* of the order *Lagomorpha*. They can be found across the globe and are renowned for their prolific breeding habits. There are seven different genera of rabbits. These include the incredibly cute cottontail rabbit of the genus *Sylvilagus* of which there are over a dozen species, the European rabbit *Oryctolagus cuniculus* and the endangered Amami rabbit *Pentalagus furnessi,* which is found almost exclusively in Japan.

Rabbits are found in copious numbers in the wild and are hardy survivors. However, their usually small size makes them ideal for keeping as pets, and in the UK literally thousands of children have a pet rabbit, which they'll normally keep in a hutch or run, in the back garden. Every day, thousands of rabbits are also sold in butchers' shops, as they're quite tasty. Either in a hutch or in a stew, the lowly rabbit will always endear itself to the British public. What the public may not take so readily to is the thought of a *giant* rabbit frightening the *merde* out of us – pardon my French – although if reports are to believed this is exactly what happened in the north east of England in 2006.

The picturesque village of Felton nestles on the north side of the River Coquet, and also sits upon the fertile coastal plain of Northumbria. It is situated ten miles north of Morpeth and roughly the same distance south of Alnwick. Its population has varied over the centuries between 500 and 800. Currently around 600 people reside there, and the village has played host to some prodigious characters over the centuries, including the Scrope, Percy and Riddell families.

Mind you, Felton has had its downs as well as ups. Back in the 13th Century the villagers took sides with the Alexander II, King of Scotland, when he had a tiff with King John. Feltonians joined the rebellion of the English barons against John, but lost. John, pissed off to the extreme, sent an army north to quell the rebels. Felton suffered severely, and was burnt to the ground. Nothing much happened around Felton after that, at least until the giant rabbit came.
There is a strong agricultural tradition in and around Felton, due in part to the fertile, loamy soil that is ideal for raising crops. Growing vegetables is an old Felton tradition. Destroying them, however, was the *modus operandi* of the giant rabbit, which fell upon this quaint little village with unbridled ferocity. The first that the world outside Felton knew of it was when headlines such as *MONSTER RABBIT STALKS UK VILLAGE* started to appear in the press.

One morning in April, the good folk woke up seemingly to find their vegetable plots at the Mouldshaugh Lane allotments laid waste in the most horrendous manner. Carrots, onions, potatoes, pumpkins, leeks and other legumes had been dug up and smashed, their remains scattered unceremoniously hither and thither. At first it seemed that a wanton act of vandalism had been responsible, although the normally sedate village rarely suffered from that sort of unwelcome behaviour. What made them particularly angry was the fact that the village vegetable show was looming on the horizon, and a large number of prize specimens had now been destroyed. Feelings were running high.

Then, one or two residents started to wonder whether vandals had been responsible at all. They noticed something that led them to believe that an altogether stranger antagonist was at work.

First of all there were the footprints. According to some reports, all over the village marks in the soil indicated that the culprit was actually a rabbit. Now rabbit feet come in varying sizes, as do those of humans, but normally fall within pretty acceptable parameters. *This* rabbit was leaving footprints large enough to convert into decent sized swimming pools. Well, I exaggerate, but they *were* bloody big.

Naturally enough, of course, many suspected that they had been the subject of a distasteful but well-executed April Fool's Day prank. But to their horror the raids continued night after night, and the miscreant responsible just couldn't be caught.

Still, the possibility that a joker was at work loomed large. Parallels were drawn between the Felton fiasco and the recently released Wallace & Gromit movie *Curse of the Were-Rabbit,* in which a giant bunny started snaffling through a hoard of prize vegetables just prior to the annual contest to determine who had grown the best leek, spud, etc. Whether the dozen or so residents of Felton who had had their vegetables pummelled beyond recognition found any humour in the similarity is not recorded, but one suspects not.

If there was any doubt that Felton was facing a real crisis precipitated by a very real rabbit, it was dispelled after residents actually caught sight of the bugger. It turned out that one resident

Above: Felton…home of a rabbit of huge proportions. **Below**: The road over the River Coquet from Felton. The arrow points to the first place where the rabbit was allegedly seen.

Felton gets busy as monster rabbit hunters
BELOW: Onlookers gather at the bridge as the hunt continues.

Near the riverbank where footprints belonging to the rabbit were allegedly found on the second day of its appearance.

The *Stags Head* public house at Felton.

The Giant Rabbit of Felton (artistic representation).

The Northumberland Arms at Thirston, just across the river. According to locals the rabbit was seen here twice.

The bridge between Thirston and Felton

A more conventional giant rabbit

Charles Dick, a revered resident of Felton, home of the giant rabbit.

had actually espied the creature back in February. He described it as "a monster" and testified that its footprints were "a lot bigger than a deer's".

"The first time I clapped eyes on it I said, 'Blimey, what the hell is that?'" he claimed. *"This is no ordinary rabbit; I think we're dealing with a bloody monster."*

Before long, the account started to spread like wildfire. Soon, it wasn't simply the rabbit's footprints that were bigger than a deer's; the rabbit was said to be "bigger than a deer" itself. More moderate voices showed restraint, and claimed merely that the beast was the size of a large dog. Still, you wouldn't want to bump into it on a darkened footpath down by the river, particularly if it was in a foul mood, drunk or feeling a bit peckish.

Within a week the creature had been given the rather impish epithet of "Bigs Bunny", and plans were being laid down to end its reign of terror. In the Wallace & Gromit film, the heroes attempt to implement what were described as "humane types of pest-control". This was far too sedate a retort for the enraged vegetable growers of Felton, some of whom advocated a "shoot to kill" policy. Trained marksmen had been "called in", the press announced, although who they were, where they were trained, and who exactly summoned them was not made clear at that juncture. Ramping up the adrenalin levels even further, the trained marksmen were later referred to as "sharpshooters", which only served to engender bizarre images of hunters in Davy Crockett hats sporting oversized blunderbusses.

The RSPCA and similar organisations also played their hand, although with restraint. Apparently they did not want Bigs Bunny shot, but merely trapped in a large cage, so that it could be released elsewhere. By this time, of course, the rabbit was supposedly the size of the Great Pyramid of Cheops, so where a suitable cage would have been found to house it is anybody's guess.

Not to be held back, the allotment holders apparently hired a guard to patrol the vegetable patches during the evening. The "guard" was later said to be a "licensed gamekeeper". Mind you, if all the accounts were to be believed there must have been precious little left to protect.

The language used to describe the creature was, after another two weeks, becoming infinitely more colourful. It was no longer simply a "monster", but also "a brute of a thing", "evil" and "Satanic". As feelings grew ever more intense, an animal welfare expert again urged restraint, asking people to, "explore all humane alternatives first", presumably before peppering it with lead. "People should employ good husbandry to keep them out", he added.

"Them?" What did he mean, *them*? My God, the Feltonians wondered, was there actually more than one?

As the days passed, descriptions of Bigs Bunny became more detailed. He (if it was a *he*) was black and tan in colour, and looked like a cross between "a rabbit and a hare". Its paws were disproportionately large, and, enigmatically, it had one ear larger than the other. This latter feature was testified to by four separate villagers who had seen the creature, and so must be given some credence.

Alongside the gamekeeper, reports talked of "two local lads" who were also determined to bring down Bigs Bunny in a hail of bullets.

"They keep trying to take a shot at it", claimed a local, *"but its damned clever"*. Considering the alleged size of its cranium, perhaps we shouldn't be surprised.
Another villager added that the local lads who were trying to hunt it down "never actually see it".

A giant, psychotic, rabbit/hare cross with the intellect of Albert Einstein and the power to render itself invisible; things were looking grim indeed. I jest, of course, but only because stories of giant bunnies lend themselves irresistibly to a touch of humour. Even when colossal rabbits are munching away at our prize vegetables and frightening the horses, we must still learn not to take life too seriously. But I do not poke fun. There is a serious side to this story, as we shall see presently.

One perceptive hack captured the mood perfectly when he spoke of "cold fear metamorphosing into vengeful anger". It was true. As the days marched on and the shortage of vegetables brought back memories of wartime rationing, the knives were out, both literally and metaphorically, for Bigs Bunny.

At an absolutely electrifying meeting one local declared, that when the animal was eventually killed it would be hung from a tree so that everyone could view it.

"We'll certainly get it" an elderly resident was quoted as saying in the press.

Within two days of the meeting, the giant rabbit of Felton exploded onto the world's consciousness. Some of the credit for turning the critter into a global celebrity must lie with the media. One on-line news site carried the headline **CURSE OF THE WERE-RABBIT**, whilst a national newspaper splattered the slightly more informative **NO LEEK OR TURNIP IS SAFE FROM THE CURSE OF THE DREADED WERE-RABBIT** across the page.

A curse, eh? So what was Bigs Bunny doing, sticking pins in effigies of carrots and turnips? The world's media wanted to know. Locals started to receive e-mails and phone calls from all over the world, as did the *Northumberland Gazette.* In the electronic missives flooded, from the USA, New Zealand, Australia and Canada. Hot on the heels of the English-speaking countries came the likes of Russia, Georgia, Nigeria, and Bahrain. One resident allegedly received no less than 82 calls in one day from all over the globe, although how the world's press got his or her telephone number I don't know. Half wanted to know the story, the other half wanted to plead for the life of Bigs Bunny. Bigsy, meanwhile – presumably unaware of the furore – just kept on wrecking the allotments in an orgy of nocturnal, sylvian destruction. Somehow, he always managed to elude his hunters.

Uproar occurred in the offices of one British national newspaper when its arch rival ran the story, **HELL RABBIT TERROR IN NORTH EAST**. A slight exaggeration, perhaps, but it almost came to fisticuffs when one reporter was vilified for not picking up the story first.

Of course, giant rabbits the size of a deer do not come ten-a-penny, so it was no big surprise that the press wanted to know what was going on up at Felton. What *did* stun the locals was the magnitude of the interest shown. One reflected that the story had *"gone right around the world. I think it's the excitement..., it's something different from listening to all the doom and gloom."*

But then again, some of the calls received were rather sinister. Another local apparently received a rather chilling threat from someone in the Midlands who said that if the rabbit was killed they would "come up there and sort you out."

Quite, but who was going to stop Bigs Bunny exterminating the locals? No matter, for new information was emerging regarding Bigsy, and it did not make for happy reading. Residents had already begun to notice that Bigs only ever seemed to prowl at night. Before long, insinuations – nay, downright accusations - were being made that Bigsy was well, *supernatural*. One chap commented, quite innocently, that he hadn't seen the rabbit in the mornings, although he did concede that it would "probably be around somewhere". Others read more into Bigsy's nocturnal habits that he did, and voiced the unthinkable thought; Bigsy may not just be a giant rabbit, but a *vampire* rabbit to boot. Oh, you may laugh, but the concept of a *vampiric bunny* is one that had already lodged itself firmly in the consciousness of locals, as we shall see later. The press didn't wait, and ramped up the fear even further with slogans such as **DEVIL RABBIT!**, **VAMPIRE RABBIT** and **FELTON WERE-RABBIT STRIKES AGAIN!** One extremely sober journal (well, usually) promptly dubbed it "THE FELTON BEAST".

The "two local lads" employed with the task of executing the Beast also became part of the legend, being given the illustrious epithets "The Hired Guns" and "The Magnificent Two". Perhaps Clint Eastwood could be wheeled out to star in a movie entitled *A Fistful of Droppings*, whilst the follow-up could be called *For a Few Droppings More*. Ye gads, these were exciting times.

An enterprising hack in the USA even suggested (tongue placed firmly in cheek, of course) that Bigsy may have crossed the Atlantic and been responsible for the theft of 200lbs of parsnips from a warehouse in Concord, Massachusetts.

As Bigs Bunny fever spread, a restaurant owner in nearby Alnwick added a new dish to his menu; *Roast Felton Rabbit Stuffed With Leeks*. At the *Stag's Head*, so I've been told, huddled groups of villagers clutching pints of ale made plans. Similar convocations allegedly took place at the *Northumberland Arms*, in the village of Thirston over the river nearby. The beast had to be stopped.

The following day, they were galvanised in their intent when a local woman saw Bigs Bunny "thundering across my garden" early in the morning. The hunt needed to be intensified.

For three nights, a 17-year-old under-keeper patrolled the allotment. To his credit he insisted that he was not frightened at the prospect of confronting the beast of Felton. He conceded that only a handful of people had actually espied the creature, but added enigmatically that he'd witnessed the evidence of the beast's existence with his own eyes, and that it was, well, "big".

He also added that he'd examined the creature's footprints and said they were similar in size to that of a large dog. Still unsettled, perhaps, he added that the sooner the creature was found the better it would be for all.

A rabbit and guinea pig rescue charity based in the south promised that they would actually journey up north and re-house Bigsy once he'd been caught. All they'd need to do was find a cargo plane or freight carrier large enough to transport him, one presumes, and the rest would be easy peasy lemon squeezy. Villagers who spotted Bigsy were urged to contact them via their website.

In a nice gesture, one animal welfare organisation offered to supply a selection of fresh vegetables to both villagers and Bigsy, in order to effect a "live and let live" solution. A spokesperson added that it seems ridiculous to demonise this animal for simply doing what it had to in order to survive, and, I would venture, this was probably the most sensible statement uttered during the entire affair. The creature *was* being demonised, and words like "demonic", Satanic", and "unholy" were being used in the press with increasing frequency. Nevertheless, one can fully sympathise with the villagers whose crops were being destroyed on a nightly basis.

We've probably laboured too much upon the activities of the beast of Felton and the effect it had upon the villagers. Other, more intriguing issues need to be addressed. What exactly was the creature? What happened to it? Is it still thumping around the allotments of Felton and the local environs?

To establish just what sort of creature the beast of Felton was, we need to look at the evidence. Essentially, this comes in two forms; forensic and eyewitness testimony. We'll look at the eyewitness testimony first.

From the testimonies of those who actually saw the creature, there is unanimous agreement about three things. Firstly, that the beast of Felton was a rabbit, and almost certainly *not* a rabbit/hare cross [1]. Secondly, the witnesses agree that it was coloured black and tan. Thirdly, they also agree that one ear was markedly bigger than the other.

---

1. Annie Gray who wrote the `bible` of Mammalian hybridisation states that there are no records of crosses between rabbits and the common European hare (*Lepus europaeus*). She states that there are a few 19th Century claims that rabbits had hybridised with the mountain hare (*Lepus timidus*) but these are very unlikely. The montain hare is not found in Tyneside anyway. Claims that such hybrids are called leporines, are a misnomer. These are actually hybrids between domestic rabbits, and a domesticated race of the rabbit called the `Belgian Hare`. Even if rabbits *had* hybridised with either species of hare, the progeny would have been almost certainly smaller, rather than larger than normal members of the rabbit population in the area.

The mystery animals of Northumberland and Tyneside

When it comes to estimating the size of the creature, there is a measure of disagreement. Subjective descriptions such as, "It's a big monster", "its absolutely massive" or "It's a huge brute", tell us little, other than that in the eyes of the witnesses the creature undoubtedly appeared unusually – nay, extremely - large. Some described it as "the biggest rabbit they'd ever seen", others "like a large dog", and yet others as "the size of a deer".

If all the reports are to be taken at face value, there is little doubt that the creature wreaked havoc throughout the village of Felton. Dozens of villagers seemingly saw the results of the animal's nocturnal pillaging.

What about the forensic evidence? Well, the creature left extremely large footprints. Gamekeepers and others well versed in the ways of the countryside and its fauna all agreed that the prints were those of an extremely large rabbit. The vegetation that was ruined by the creature also bore tell-tale signs. Huge bites were taken out of turnips, leeks and other vegetables.

One witness bemoaned the fact that they hadn't had much luck catching it, whilst another exclaimed, "*My God, have you seen what its done to the leeks and turnips?*"

Again, these bite-marks allegedly bore all the classic hallmarks of having been made by an extremely large rabbit. We also need to bear in mind the colossal amount of vegetable matter ingested by the beast. The creature was so powerful it regularly pulled prize leeks from the soil - and on more than one occasion allegedly ate *entire rows* of leeks onions, carrots and parsnips. On one foray it was said to have ingested "a market stall's worth" of vegetables.

We now have to address the obvious question; what indeed was or is the beast of Felton? We can safely say, I think that it was definitely a rabbit or, as some have suggested, some sort of rabbit/hare cross. The problem with the rabbit/hare cross theory is that fertilising rabbits with hare spermatozoa is extremely difficult. If the beast of Felton was such a cross, then it must have been an extremely unusual one. It is far more likely that the creature was, despite its unusual size, a rabbit plain and simple. But what sort of rabbit?

Initial press reports quoted an "expert" as saying that the beast could have been an escaped pet rabbit of a "giant" breed. The current record holder for being "the world's largest bunny" is actually found in England, a Continental giant named Roberto. According to reports he weighs 15.9 kilos and is 107 centimetres long.

Experts concede that some rabbits do indeed grow to huge proportions. Continental giants can occasionally exceed 26 inches in length. Other experts suggested that the creature may have been a Flemish giant, which, allegedly, can grow to "three feet in *height*". Presuming that this is not a typographical error, one dreads to think of what the *length* of such a specimen might be.

Some locals offered another explanation, although it was in some respects problematical. They argued that back in the 1950s and early 1960s, before the introduction of pesticides, rabbits commonly grew to much larger proportions than those now seen in the countryside. This may be true, but given that pesticides are now almost universally present in the countryside anyhow, how does one explain the size of a rabbit like the beast of Felton?

There are no easy answers. All we can say with certainty is that the creature was a rabbit and, by all accounts, absolutely enormous.

A superficial examination of the evidence would suggest that the fate of the beast of Felton is easy to determine. One evening, an 18 year-old A-Level student was driving on an approach road to the A1 when she ran over a "massive" rabbit with her car. The impact fractured the bumper of her vehicle and a large tuft of fur was found inserted in the tear. The woman left her vehicle and went to examine the damage. There, stretched out on the road, was a huge rabbit. The beast of Felton had, it seemed, been sent into the Great Beyond.

Or perhaps not. One of the interesting aspects of her account – and we have no reason to doubt it – is that the creature she killed was "about two feet long". This is large, but in no way measures up to the dimensions of the beast of Felton. Even a rabbit of this length wouldn't be able to ingest the huge amount of vegetation that the beast could devour.

After the accident, a trickle of sightings still came in which indicated that the beast of Felton was still alive and well. True, the ravaging of the village's allotments ceased at the same time, but one wonders if this may just have been coincidence.

Is it possible that the beast of Felton may have been a "zooform" creature; that is, not a creature of flesh and blood, but something altogether more spectral in nature? It's interesting that one media report described the creature's ears as "diabolical" in shape. One resident I spoke to said that he'd seen the beast "pretty close up", and that he found its ears to be "unsettling" when he looked at them. "There was something about those ears. I don't mind telling you that they scared me. I can't explain why."

I've found that this is a common them with those who see "zooform" animals; one particular feature of the animal seems to engender strange sensations and feelings of dread in the experient, and yet for no rational reason. We may have just a hint here, then, that the beast of Felton may have been altogether stranger than a mere giant rabbit.

In August 2007 my wife and I visited Felton and quizzed some of the local residents about the affair. One elderly lady looked at us askance and said, "*It was a hoax!*"

Really? And who, pray, would have engineered such a spectacular prank?

"*It was the government! Mark my words! They did it just to get publicity!*"

This was highly unlikely, of course. The woman was merely expressing a totally unworkable opinion in the absence of any sensible ones. It was obvious that she could not countenance the idea that the giant rabbit of Felton was *real*. Like all good cryptozoological animals, the beast of Felton seems to have disappeared. Feltonians are once again able to organise their annual vegetable show unhindered and the allotments no longer need to be guarded against this fearsome intruder.

There is a strange postscript to this story which, although not tied directly to the appearance of the Giant Rabbit, presents itself as one of those weird coincidences which Fortean researches seem to attract like magnets.

Back in the 1960s, when I was a young'un, I often used to visit my paternal grandparents on a weekend. In fact, they only lived several streets away. My grandfather, known in the family as `Granda Bill`, was a man of moderate habits. He drank only rarely and his only vice was tobacco. As I recall, he used to smoke around 60 Players un-tipped a day. Now, of course, you can be jailed for life merely for lustfully looking at a packet of cigarettes never mind smoking them, and tobacco use is fast becoming demonised to the same degree as murder, terrorism and morris dancing. But it was different back then, and I have pleasant memories of seeing my grandfather sitting in his favourite armchair with three things; a copy of the *Daily Sketch (*always opened at the racing pages). His cigarettes and ashtray and, crucially, a mug of tea (it had to be CO-OP tea – nothing else was allowed in the house). My grandfather liked to have a weekly flutter on the horses. He never spent much, but he seemed to win quite a bit. *Why* he was so successful is where the story gets interesting.

About twice a year, and always on a Saturday afternoon, a man used to visit my grandfather. He was, as I recall, quite tall and always wore a finely-checked sports jacket and grey slacks. I also remember distinctly that he sported a trilby which was pushed back on his head in a manner later popularised by Arthur Daley in *Minder.* I never knew the name of the man, but I recall that my grandparents always spoke well of him. Even at a young age I remember thinking that he was a gentleman, and on occasion he's slip me two shillings with a wink and a smile. For an hour or two my grandfather and the man would chew the metaphorical cud, putting the world right and chatting about horses. He'd give my grandfather tips on the horses, and they never failed. His accuracy was stunning, and no one in the family could work out how he did it. Eventually, however, the the visits of this man tailed off, and I totally forgot about him.

Fast forward to the year 2002. I was, then, chairman of the Twilight Worlds Paranormal Research Society, and I recall a visiting speaker arriving from Northumberland. With him was his good wife and an elderly friend by the name of Charles Dick. My wife Jackie and I immediately hit it off with Charles, and he'd had an extremely interesting life. He was a skilled wood-turner and "countryman", and also evinced a genuine interest in paranormal research and other esoteric subjects, as well as writing poetry. Back in 1946, the Northumberland & Alnwick Gazette published a book written by Charlie entitled, *The Poems of a Prisoner of War.*

I remember having a grand old chat with Charlie about man-beasts, anomalous big cats and other cryptozoological matters. He was never quick to express an opinion, but what he did say always made sense. He was, quite simply, fascinating to talk to. Charles Dick and I corresponded for a while, and I will always retain fond recollections of the times we spoke. I never did find out where Charles came from exactly, although I remember him saying that he lived in Northumberland. Later, I received the news that he had passed away with great sadness.

When I was researching the Giant Rabbit story and visited Felton to interview some of the locals, I was amazed to see a plaque dedicated to the memory of Charles Dick on a wall. So *that's* where he'd came from! If only Charles had been alive when the Giant Rabbit story had broke he'd have been able to give me the true story without a doubt. Fancy stumbling on something like that in a remote Northumbrian village like Felton, I mused.

In March 2008, I happened to mention to my father that when I'd been at Felton I'd found a plaque dedicated to an old friend, and had been intrigued by such a weird coincidence. What happened next was even weirder.

"It's funny, you know; your grandfather used to have a friend from that neck of the woods. He used to come about twice a year and give him tips on the horses".

Suddenly the memories came flooding back, and I recalled the old chap my father was speaking about.

"My God...I remember him. He used to wear a checked jacket and a trilby...am I right?"

"That's the man", my father replied. *"Like I say, he came from up that way somewhere".*

"What was his name? I was only young at the time, and I don't recall it".

"He was called Dick*...that was his surname".*

"It wasn't Charles Dick, was it?"

"Yeah, that's it; Charles Dick".

Theoretically, its just possible that the Charles Dick who visited my grandfather all those years ago and the Charles Dick who I met so much later in life were two different people, but I don't think so. I don't recall the Charles Dick I met at Twilight Worlds ever expressing an interest in horse-racing, but then again, we had more important things to talk about. I just found it incredibly strange that our paths should cross again after such a long time, and yet neither of us knew it. As I say, its the sort of bizarre coincidence that Forteans are used to. I don't know why they occur, but Charlie, if he'd still been around, might have been the guy to ask.

Was the beast of Felton a one-off mutant freak, or are there others out there lurking in the Northumberland countryside? Did the beast ever exist in the truest sense, or were the descriptions of its colossal size merely generated by over-excited imaginations? We may never know.

Chapter Two

The Ghost Birds of Jesmond Dene

When one lives in an intensely urbanised and industrialised area like Tyneside, as I do, it is all too easy to become disconnected from the past. The relatively *modern* past isn't so much of a problem, as urbanised areas normally play host to numerous museums, which detail the life and times of past residents. A trip to the local library will also inform the curious traveller of numerous buildings and sites that can be visited if one wishes to find out about the agricultural, industrial, political or cultural life of previous generations. The ancient past – the *truly* ancient past – is a different kettle of fish altogether. If one stands in a modern city centre, filled with shopping malls and community centres, one gets little or no sense of how the landscape may have appeared millions of years ago. Modern life forms a metaphorical fire blanket, which effectively insulates us from bygone ages when dinosaurs walked the earth and the flora and fauna was radically different to what we see now. Nevertheless, look hard enough and it's still possible to detect faint echoes of those times. Jesmond Dene, in Newcastle upon Tyne, is one such locality.

310 million years ago, the entire area around Jesmond Dene was a huge delta flowing into an ocean roughly where the North Sea is now. Over the millennia, sandy deposits were deposited by the slow-moving currents, and a veritable spider's web of channels was formed. Eventually the waters in the delta subsided to a degree, and that which was left became frozen with the onset of another Ice Age. As the Ice Age passed its peak, the frozen water receded. As it retracted, it cut huge gouges into the rock below, adding to the already complex maze of channels and distributaries. In this powerful, but painstakingly slow process, Jesmond Dene was formed; a glacial stream forming part of the

connection between the Ouseburn and the River Tyne.

As the Ice Age receded further, the climate warmed and the variety of flora became ever more diverse. Meagre clumps of moss were eventually replaced with huge forests. These forests provided a home for many of the animal species, which are now extinct – although as a true cryptozoologist I should perhaps say *probably* extinct. One can never be absolutely sure, don't you know?

In time, nature would give birth to the oak, lime, hazel, birch, holly, beech and ash trees. These species gained a foothold in what would later become known as the north of England, and most of them can still be found growing at Jesmond Dene. Are they the direct descendants of arboreal ancestors who inhabited the same site millions of years ago? We can't be sure, but it's a fascinating thought.

In the mid 19th Century, Baron George Armstrong owned much of the land around Jesmond Dene and was responsible for building a number of illustrious structures, including an imposing Banqueting Hall. A great lover of nature, he also had imported a wide variety of exotic shrubs and other plants, many of which still thrive in the area even now. In the 1880s, Armstrong presented the Dene to the Mayor and Aldermen of the city of Newcastle upon Tyne as a public park so that local people could enjoy the fruits of his labours. We should thank him for his generosity. The then Prince and Princess of Wales formally opened the park, and to mark the event the princess planted a young turkey oak near the Banqueting Hall. Now a large and sturdy specimen, the tree can still be seen in the park to this day.

Not long after Armstrong bequeathed the park to the people, swans began to nest on the Dene itself. These proved to be a particular attraction to visitors and gained an almost royal status. They were never there in large numbers, but their majesty and grace proved to be an inspiration. Apparently, the focal character amongst the swan population was a jet-black male who drew much attention from human visitors, and revelled in it.

Adjacent to the Dene is a small, grassy area, which contains a number of diminutive gravestones. True these memorials remind us of our mortality, but there is something serene about this burial ground. It is as if the river itself soothes the souls of those buried beside it as it tinkles past their final resting places. The notion that there may be a strange relationship between these memorials and the swans that once lived nearby is one that demands examination, but we must not get ahead of ourselves. First we must look at a number of bizarre experiences that people have had near the Dene itself.

Not far from the entrance to Jesmond Dene the visitor will see Pets' Corner; a small sanctuary which currently houses an exotic pig, some pygmy goats, and a beautiful array of flamboyant birds including a spectacular pair of peacocks.

But, what of the swans? Alas, by 1990 they were no longer to be found on the Dene. According to some they were either killed or driven off by mindless hooligans, and, sadly, they're probably right. Now, it seems, the only swans to grace the skies over Jesmond Dene may be of

Jesmond Dene – once home to the enigmatic Ghost Swans which may still haunt the area.

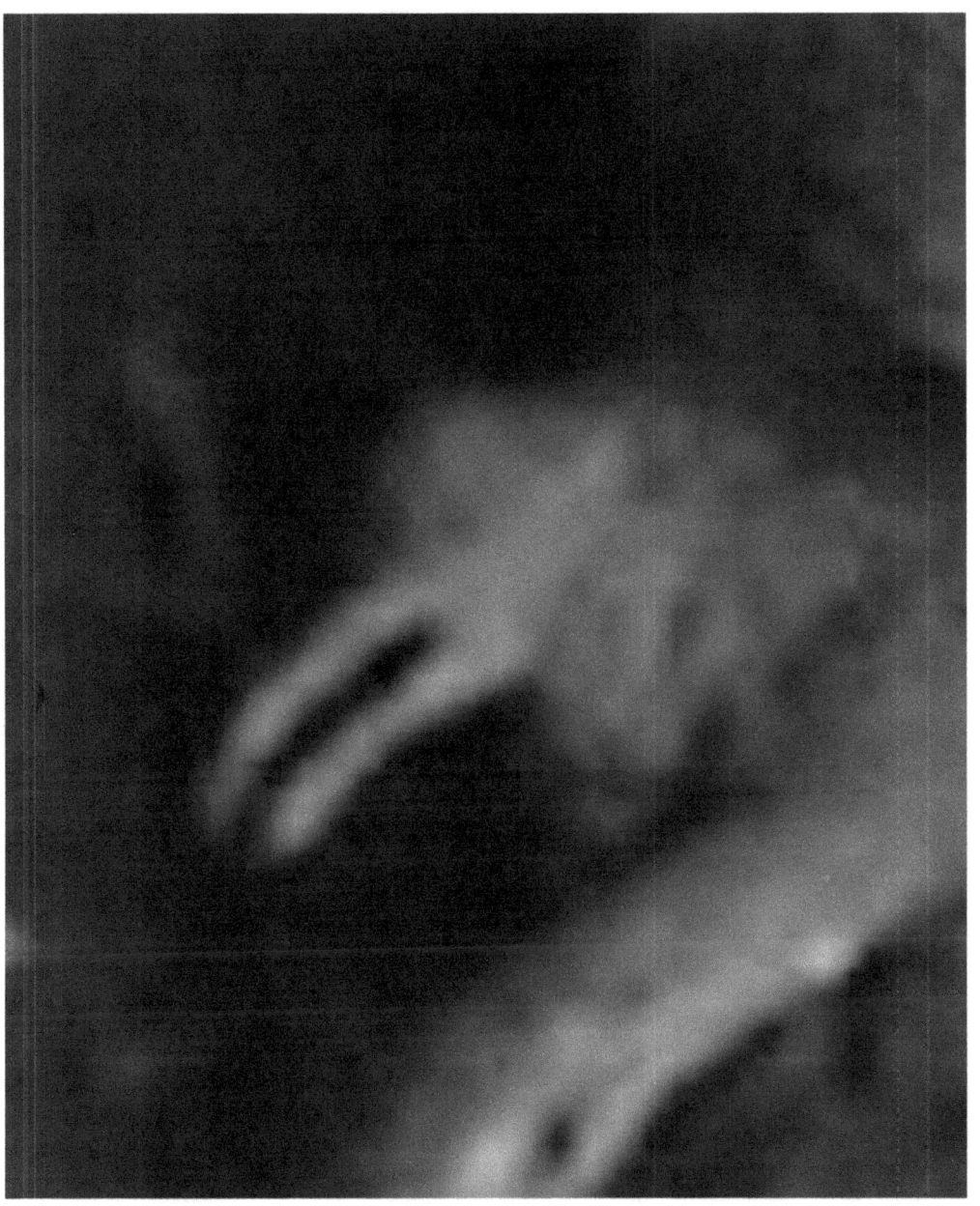

Three bird simulacra

The graves that line Jesmond Dene – the spirits of the dead allegedly take the form of swans before flying overhead.

Jesmond Dene

the spectral kind.

Strip away the buildings and new features, and the timelessness of Jesmond Dene quickly bubbles to the surface. The park is immaculate, but cultivation is left to Mother Nature wherever appropriate. Walk down into the valley and civilisation soon falls sharply into the background. There is a "wilderness" feel about much of Jesmond Dene, even though you know that the 21st Century is simply lurking around the next corner.

Jackie and I visited the areas of the river where the swans used to rest, and it is true indeed that one can almost sense their presence even though they are no longer there. You know that they *should* be there, but they just aren't. The connection between birds (and in particular swans) and death is something that goes back to the time of Shakespeare and beyond, and Jackie and I had heard persistent rumours that spectral or "ghost" birds (usually swans) had been seen on numerous occasions at Jesmond Dene. We went there, took photographs, and hoped that some would put in an appearance, but alas it was not to be.

On our second visit to Jesmond Dene I interviewed a number of visitors and asked them a) whether they had heard of the ghost birds and/or b) if they had seen them. Most hadn't, but a small number answered in the affirmative. Tourist Tony Smith said he'd *"read about them in a book years ago"* although he couldn't remember which one.

"I've heard about them, yes...but I've never seen them. They're supposed to be the spirits of the dead or something. My aunt lives in Kenton, and she said she knew someone who claimed to have seen them, but I don't know; maybe she just read the same book as me and her imagination got the better of her!"

Tony couldn't give any details about the sighting of the ghost birds allegedly made by his aunt's friend, so one supposes that the jury is still out, as they say.

Ethel – who didn't want her surname or address immortalised in print - said that she may indeed have seen the ghost birds, but wasn't sure. She claimed that in 2002, when she was a sprightly 67 year-old, she'd brought her great-grandson to the park for a walk. She could not recall exactly when, but believed that it was "during the school holidays", which means the incident likely took place in the month of August.

"We were walking along the river, when we suddenly heard a fluttering noise in the trees to our left. I turned and looked, but couldn't see anything. Mark said, 'What was that, Grandma?' I honestly didn't know what it was, so we walked on. A minute or two later we heard it again, but this time I looked up and for a brief second I saw some things fluttering above my head. I couldn't make out what they were, because for some reason I couldn't seem to make out their shape, but they were something like big birds. It was all very strange, and I started to feel a bit uncomfortable. Mark and I quickly walked back up the path the way we had come.

Later I mentioned what I'd seen to my husband, and he said that they had probably been bats.

The thing was I've seen bats before, and these didn't seem like bats to me. I told my husband that bats only came out like that at night, but he insisted that it was probably bats I'd seen. I still don't think so."

The insubstantial nature of the "things" seen by Ethel is intriguing, but if the witnessing of the ghost birds is fascinating in itself, working out exactly what type of phenomenon they may be is even more so. Superficially, one could argue that, if they exist, the spectral birds are as their epithet suggests; ghosts. Animal ghosts are not as common as their human equivalents, but they certainly aren't unknown. A well-known public house in South Shields allegedly has at least one (and probably two) ghost cats. One, which normally inhabits the front bar, used to brush against the legs of unwitting patrons busy playing the one-arm bandits.

What complicates the issue somewhat, at least insofar as the ghost birds are concerned, is the presence of the aforementioned graves near the riverbank. Do the souls of the dead transmigrate into swans or ducks before flying off into the world beyond? A romantic notion, I would venture, but one with a historical pedigree.

A fellow researcher told me that there are birdlike simulacra all over Jesmond Dene, and its true. Study photographs taken in the area and you'll soon see what I mean. I've reproduced several in this book to give you some idea what I'm talking about.

The truth is that the ghost birds of Jesmond Dene do not fit comfortably into any conventional cryptozoological niche. It certainly seems that the ghost birds exist in some shape or form, but we simply don't have enough evidence to say with any certainty what they are or what sort of phenomenon they represent. Until such evidence is forthcoming, perhaps we should just enjoy their presence and wonder at them, giving thanks for yet another enigma that enriches our lives so wonderfully.

Chapter Three

Black Dogs & Howling Dogs

Before we turn our attention to black dog stories from the north east of England, we need to digress slightly and review a classic black dog tale from further south which will give the reader who is new to this subject a good idea of the sort of beast we're dealing with.

Churches are normally very safe places to be, and the greatest danger therein probably lies in getting a severe pain in your wallet when the collection plate is passed around. Sadly, attending church proved to be a rather less-than-pleasant experience for two parishioners in 1577. In fact, it was the last thing they ever did, and should prove to those who might doubt it that zooform animals can be every bit as deadly as their flesh-and-bone counterparts.

The black dog may quite literally terrify the life out of people. Several witnesses - and I'm being serious here - have actually gone insane after their encounter. One farmhand from Lancashire described his encounter thus: *"It was the biggest dog I ever saw. It was covered in matted, black hair and it stunk to high heaven. It had two, glowing red eyes and horrible teeth. It growled at me and seemed to grow in size as I stared at it* [another oft-reported feature of zooform animals], *and I just ran for my life."*

Enigmatically, a small number of witnesses claim that the black dog was actually friendly, and walked beside them in a protective manner to keep them safe from harm. Maybe they caught Auld Shuck on a good day, unlike the two aforementioned worshippers from the Parish of Bongay (now Bungay), in Suffolk. The story is reasonably well documented - for the time - and appeared in a pamphlet entitled *A Straunge Won-*

der In Bongay. [1] The account was composed from the written testimony of one Abraham Fleming. What follows is a reconstruction of the event, largely (but not exclusively) based on Fleming's observations. I have dramatised the account slightly, simply to allow me to create a coherent sequence of events, as Fleming's account is, despite its likely accuracy, annoyingly brief.

It was early in the morning on Sunday, August 4, 1577. The weather was absolutely atrocious, with the rain pouring down in torrents and arcs of lightning shooting across the sky. The thunderclaps were so loud they seemed to shake the little Bongay church to its very foundations. Nevertheless, the parishioners had turned out as usual for their devotions and to hear a sermon. Inside, the congregation knelt in prayer, their heads touching the wood of the pew in front and their hands clasped together. Not in their worst nightmares could they have imagined what was to happen next.

Quietly, the door at the back of the church opened; an event which would have passed unnoticed by the parishioners due to the incessant thunderclaps which masked all other sounds - including most of the sermon, in all probability. The Vicar was oblivious to the door opening too, for although he faced the entrance from the pulpit, his head was bowed and his eyes were shut as he led the flock in supplication. Then, through the door, walked a huge, black dog, the likes of which the people of Bongay had never seen before. It was almost the size of a small pony, and covered in matted, black hair. It had glowing, red eyes and large paws. It was hideous to look at, and exuded an aura of sheer menace. It also reeked of sulphur. The creature paused momentarily, gazed at the assembled devotees, and then proceeded up the narrow aisle.

Two parishioners - both male - occupied the seats nearest the central aisle in the last two pews to be occupied towards the rear of the church. One occupied the inner left-hand facing seat, the other the inner right-hand facing seat. Had they stretched out their hands across the aisle they could have touched each other with ease.

It is unlikely that these two men ever knew what hit them. Several people had now become aware of the dog's presence, and turned in amazement to stare at its hideous form. They focussed their gaze just in time to see the beast walk between the two praying men. Without warning, the dog reared up and brought it's huge paws down; one on the neck of the parishioner on the left, the other upon the neck of his friend sitting across the aisle on the right. The other worshipers looked on in disbelief as the animal, in the words of Abraham Fleming, *"wrung the necks of them bothe at one instant clene backward, insomuch that even at a moment where they kneeled, they strangely dyed."*

At this point, if you will excuse the malapropism, all hell broke loose. Sitting nearby was an-

---

1. The title in full is: *A straunge, and terrible Wunder wrought very late in the parish church of Bongay, a Town of no great distance from the citie of Norwich, namely the fourth of this August, in ye yeere of our Lord 1577. In a great tempest of violent raine, lightning, and thunder, the like wherof hath been seldome seene. With the appearance of an horrible shaped thing, sensibly perceived of the people then and there assembled. Drawen into a plaine method according to the written copy.* London, 1577.

other male parishioner, who bravely lunged at the beast in an effort to wrestle it to the floor. The dog merely lashed out with its paw and caught the man on his back, felling him instantly. Fleming would later say of this man, "*that therewith all he was presently drawn togither and shrunk up, as it were a piece of lether scorched in a hot fire; or as the mouth of a purse or bag drawen togither with a string. The man, albeit hee was in so straunge a taking, dyed not, but as it is thought is yet alive."* Lucky him.

Having wreaked havoc upon Bongay church, the dog silently plodded out of the door as enigmatically as he had entered it. The town of Blibery was just a few miles up the road.

When the good people of Blibery [now Blythburgh] assembled in their church for the Sabbath Service, they were unaware of the events, which had transpired earlier at Bongay. In good humour, the farmers, merchants and blacksmiths all trooped in with their families to thank the Almighty for the week's blessings. As the parson took his place, the chattering stopped. The parishioners wanted the spirit of God to come down upon them. What they got was a colossal black dog, which had been perched on an oak beam above the congregation. As the huge, snarling canine propelled itself through the air - how he actually got onto the beam was never satisfactorily explained - the parishioners jumped round in astonishment. The first victim never even got that far, his neck snapping instantly as the creature landed upon him. The man next to him fared no better, the dog clipping him across the head with its paw. He was dead before he hit the ground.

Pandemonium broke out. Terrified parishioners made for the door, fathers and mothers snatching their youngsters to safety as the dog chased after them. One young boy, frantically looking around for his parents, caught the full weight of the dog upon his back. He died instantly.

Bravely, one woman pushed at the dog with her hand, only to find, to her horror, that the flesh was instantly incinerated. She had been hideously burnt simply by touching the creature. Now the dog was in full flight, leaping from pew to pew, knocking aside fully-grown men like rag dolls. He was burning them, breaking their bones. Suddenly, it tired. With an almost stately air, the hideous animal padded down the aisle of Blibery church and exited via the large oaken doors. Once outside it paused, momentarily, and turned. Lifting one of its huge paws, it hacked at the door, leaving a series of deeply ingrained scratch marks in the wood. Parishioners would notice later that the scratches had actually been *burned* into the wood, as if branded there with a hot iron. Black dog then sauntered off, never to be seen again, at least under such horrific circumstances. The church still stands at Blythburgh, and the original doors are still in place. So are the scratch marks; a testimony to a sceptical world that, even if we are dumb enough to believe only in what we perceive as rational, our ancestors were under no such restrictions.

It has also been suggested that black dogs often appear on the paths of ley lines, and as churches are often found on ley lines then the presence of the black dog at Bongay and Blibery may support such an assumption. On the other hand, black dogs are usually seen at dusk or after dark. The presence of this prize specimen during the morning, then, is somewhat unusual.

The black dogs that frequent Northumberland are pretty much the same as those found elsewhere in the UK. Although spectral canines of other colours have been reported, black is the usual order of the day. Their coats are often described as "scruffy", "matted" or "shaggy", as if in need of grooming, and they are occasionally said to be headless. There is a curious parallel here with human ghosts, which are also sometimes seen without a head. This may just be coincidence, however, so perhaps we shouldn't make too much out of it.

Generally speaking, black dogs are described as being much larger than normal canines. Some have been estimated to be roughly the same size as a cow, and it seems to this author that a prize fight between a black dog and the giant rabbit of felton would be an interesting spectacle indeed.

Zooform animals are often described as having glowing eyes, usually red in colour. When I investigated the curious case of the 'Cleadon Panther' – more of this later – several witnesses reported the same, strange characteristic of the beast they encountered. Some witnesses have reported that the eyes of black dogs are also hideously large - "like tea plates" – although those who saw the Cleadon Panther reported nothing like this at all. The large eyes are a phenomenon more often associated with black dogs than any other spectral animal.

In general terms, black dogs can be said to be distinctly off-putting in appearance. Foul-smelling saliva will drool from their lips, their posture continually makes witnesses feel that they are about to attack them, even though they almost never do. Several experients have claimed that the black dog they encountered actually smiled at them, although even then the expression wasn't so much a smile as a menacing, evil grin.

During a trip to the USA in 2003, I stopped over briefly in Memphis. My flight to Dallas Fort Worth was delayed, and I was waiting for the people from Northwest Airlines make an announcement. I sat down in the departure lounge and was instantly drawn to a man who appeared to be talking loudly to himself. He wasn't, of course. A wire trailed from his ear to his mobile phone. He was speaking "hands free" to someone at the other end.

"Well...it all depends how we format this, Gary; I mean, he published two years ago and the damn thing went belly-up, remember? Didn't I tell you it would belly-up, huh? Didn't I get that right? Yeah, I got that right. Huh?....yeah....yeah....I know, too risky...but if we could convince their people to transfer...yeah....,yeah....oh really?...at their New York office?...You know, I was saying to Carl that we could get that contract...Lemme tell you, Gary.....that company is going places..."

I took a long, hard look at the man. I recognised his type. He was the sort who had a calculator welded permanently to his hand and whose idea of a good night out was to work late at the office. Not wanting to know any more about Gary, Carl or his effing New York office, I looked around for somewhere else to sit. As if by magic, I noticed a bar – *Charlie's Place* - opposite the departure lounge. I found a seat, ordered a Jim Beam on the rocks from the barman, and tried to relax. There was a young, black woman standing at the other end of the bar. I couldn't figure out if she worked there or was just a friend of the bar tender, but she turned and looks at me.

"Hey, why so glum?"

"Well, my flight is delayed and I've got friends waiting for me at Fort Worth. I've tried to use my mobile phone but I can't get a network connection, and I've tried to use a payphone but they don't answer".

In her hand she had her own mobile phone. She held it forward.

"Here...use mine".

"Well, that's really kind, but if they didn't answer when I got through on the payphone I guess they won't answer whatever phone I ring from. Thanks anyway, it was kind of you to offer".

"That's okay!" she chirped sweetly. It's funny how little random acts of kindness can inspire you just when you need them.

A fellow commuter - Gerry - sat down on a stool beside me. He was blond, grey-suited and a tad dishevelled. He carried a battered black brief case. He also looked as if he could do with a good night's sleep. Something told me instinctively that he was a salesman. We struck up a conversation, and he told me an eerie story about his grandmother and an encounter she'd had with a strange, other-worldly creature known as a "graveyard dog". I made a mental note of the details, and decide to write the story up for my *WraithScape* column when I got home.

Gerry's grandmother had had an encounter with a transatlantic version of the black dog; the *Graveyard Dawg*. Graveyard Dogs are said to be the spectres of dogs buried in cemeteries to protect the souls of the dead. Some say they were occasionally buried alive. According to Gerry, the *dawg* encountered by his grandmother had "huge, long teeth like needles." Long, razor-sharp teeth is a feature often noted in black dogs seen within the British Isles. Such cross-correspondences are intriguing from a researcher's point of view, as they hint at some sort of relationship (cultural and/or physical) between phenomena separated by vast geographical distances. The several tales of black dogs from the Northumberland area that I have collected contain three in which the witnesses spoke of the beasts possessing unusually long and sharp teeth.

There seems to be two distinct kinds of black dog; those who make a noise and those who are silent. Those who make a noise may bark, or more commonly howl eerily. Some make strange screeching or cawing noises more like a bird than a dog, whilst others make that horrid cross between a howl and a wail common in foxes. More rarely, some black dogs have been known to speak, although none to my knowledge have demonstrated a witty repartee or desire to discuss string theory, Marxist politics or the advent of the Industrial Revolution.

During my research I've came across few sightings in which the experient actually *saw* the black dog appear. One minute the experient is unaware of it, the next the black dog is there in full view. There is an intriguing parallel between this phenomenon and that known as *lithobolia*. Lithobolia is the technical term for a phenomenon closely related to poltergeist activity. Witnesses often report stones being mysteriously thrown both indoors and outdoors by invisible assailants. The peculiar thing is that they rarely or never see the stones appear or *begin* their movement. One moment they are not there, the next they are; however, the transitional phase between these two states is never witnessed. The parallel between this and the appearances of black dogs is intriguing. As the two phenomena are markedly different, we must be careful about forging links where none truly exist. However, it just may be that there is a common denominator present in both types of apparition, and we would do well to bear this in mind. In one case, which I'll relate presently, both the black dog and the poltergeist phenomenon of lithobolia were present simultaneously.

Most experients say that, typically, they'll be walking down a remote road or path when, suddenly, they will become aware that a black dog is walking beside them. They will not see it appear; merely become aware of its presence. Intriguingly, the *opposite* is true when the black dog disappears. Many experients recall that the creature simply "faded from view", often very slowly. Others record that the spectre disappeared in a sudden flash of light or a puff of smoke. A small number have stated that the animal's lower limbs seemed to melt away, and the beast slowly but steadily sunk into the earth until it was no longer visible. What all this means from a scientific point of view heaven knows, but its interesting nonetheless.

One of the curious features of the black dog enigma is that very strange things indeed seem to happen when experients attempt to touch the creature. There is a strong tradition that touching a black dog – I imagine few souls would actually be brave enough to try – can be fatal. Several tales record that those who made physical contact with the creature were hideously burnt. The aforementioned experiences at both Bongay and Blibery at least put forward a *prima facie* case that physical contact is extremely dangerous.

In recent years, a number of researchers have argued that black dogs have been given a bad press unnecessarily. They are essentially harmless, it is argued, and may even appear as protectors. Of course, we cannot know what the purpose is behind their appearing, and without that knowledge it is hard to quantify the degree of risk present during an encounter. However, there are enough anecdotal tales in circulation to suggest that, whatever purpose they serve, they should be treated with caution and not approached.

Of course, one's interpretation of the black dog phenomenon will largely be governed by one's cultural background and/or spiritual orientation. A Roman Catholic will interpret the experience radically different to a Choctaw Indian, and someone from Hexham in Northumberland will hardly see a Black Dog in the same context as a Kalahari Bushman. Some see black dogs in a similar light to Banshees; as harbingers of death. Others see them as spirit guides or even human souls encapsulated in animal form.

A more scientific explanation is that there may be an electromagnetic cause for the phenomenon. Black dogs often appear during thunderstorms and lightning strikes. Heavy discharges of electromagnetic energy may affect the human brain in ways that generate bizarre images. At

The mystery animals of Northumberland and Tyneside

Bongay and Blibery there was certainly thunder and lightning; but how one can account for the death and destruction visited upon those two churches is hard to explain. Were the victims killed by lightning strikes, during which witnesses hallucinated that they saw a black dog? Possible, perhaps, but I think it extremely unlikely.

Whatever black dogs are, they are a truly curious phenomenon, and Northumberland and Tyneside have witnessed their fair share of encounters over the centuries. Some have been seen at sites of historical interest, such as Chillingham Castle and Tynemouth Castle, although their appearances seem to be infrequent.

Cragside is a delightful country house which lies in the heart of the Northumberland countryside, and bears the distinction of being the first abode in the United Kingdom to be lit by hydroelectric energy. The house nestles in a craggy hillside above a picturesque arboreal garden. Formerly the home of Lord Armstrong, it has been looked after by the National Trust for over thirty years. The house takes its name from Cragend (or Crag End) Hill which sits at a higher elevation. Built in 1863, it was initially quite a modest residence, but it was later extended by the famous Victorian architect Richard Norman Shaw who some hail as the greatest designer of his time. Shaw, a Scotsman (1831 – 1913), transformed Cragside into an elaborate mock-Tudor mansion, which still delights visitors to this day. Even in those days there was an air of mystery about Cragside. In its heyday it contained a sophisticated astronomical observatory and a scientific laboratory. Study Cragside after dark, and it's not too difficult to imagine a Dr. Frankenstein or two mixing sinister chemicals together in the basement.

Turn away from Cragside and it's very easy to forget that the place exists, for it sits amid a small wilderness that effectively cuts it off from the nearest hives of civilisation. Cragside, you see, is surrounded by one of the world's largest rock gardens. This veritable Eden is filled with beautiful coniferous trees, one of which – a monstrous Douglas Fir over 60 feet in height – is the largest tree in the United Kingdom. Cragside is said to be haunted, although I honestly don't know whether this is true. What I do know is that, in times past, the surrounding environs were sometimes visited by black dogs.

There is a particular type of black dog often reported in Durham, North Yorkshire and Northumberland. Quite often they are referred to as the Gabriel Hounds. Unlike other black dogs, which may appear essentially canine from head to foot, Gabriel hounds are markedly different in two respects. Firstly, they are said to possess human heads bearing hideous physiognomies. Secondly, they do not walk, but fly. Jeff Newton told me that his grandfather, William, had been walking near Cragside one evening in 1893 when he witnessed a pack of Gabriel hounds flying overhead. *"He really didn't like to talk about it"*, he said, *"as he'd found the whole experience very frightening. Every now and then, though, he'd drop a few details if he was in the mood"*.

According to Jeff, his grandfather had heard an eerie "cawking" noise in the distance. At first he thought it was a flight of birds, but he was curious, as their cry did not correspond to any species he knew of. As the sounds drew nearer, the man craned his neck and looked skyward. To his horror he saw six or possibly seven black dogs soaring through the heavens approximately one hundred feet from the ground. *"At first he just stood still, hoping they would pass, but one of them happened to glance down upon him. Grandfather swore it had a human face. Then...then he just ran for his life. He never visited that area again – ever."*

According to William Henderson[1], William's experience was a rare occurrence indeed. Although of terrible appearance, Henderson states that many people had heard the Gabriel hounds, but few had actually seen them.

In some accounts, the creatures have the body of a swan and the head of a snarling hound. Witness Graham Randle claims to have seen the Gabriel hounds in 1992, and said they were "like swans, but with very long necks...and they had dogs' heads. There must have been five or six of them...they passed over quick so I couldn't be sure. The heads looked like they were barking, but the sound wasn't like a bark. It was more like a squishing, slobbering noise".

Randle has refused to go back to the area again, although he did point out to me the exact spot where he had his encounter.

In Northumberland, the notion that Gabriel hounds were or are harbingers of death is common among folklorists. Should the hounds pass over your dwelling, then death may well follow within the hour.

Of course, the reader will immediately be drawn to the parallel between the above-mentioned sighting of the hounds and the encounters with the Jesmond Dene ghost birds detailed earlier. Both creatures fly in groups over or through forested areas, and both make eerie cries. Further, both are said to be closely connected with the transportation of the souls of the dead. In fact, the only meaningful difference between the two (and it is admittedly a large one) is the physical appearance. The ghost birds are spectral *birds*, the Gabriel hounds are usually dogs with human heads.

But even this difference may be explainable.

Jeff Newton's grandfather, William, saw his pack of Gabriel hounds "one evening". In fact, black dogs of all types seem to prefer the hours of darkness in which to appear. Of course, a distinct lack of light raises the possibility that the witnesses may not be seeing the creatures exactly as they are, but only, in the fading light, as they have been culturally trained to see them. If you believe in ghost swans, then that's what you'll see. If you fear the dreaded black dogs, then that's what your mind may imagine.

Mind you, not for one moment am I suggesting that the tales related here are purely imaginary; not at all. I genuinely believe that the witnesses did see something extraordinary. All I am suggesting is that the two phenomena may be more closely related than at first seems likely. Henderson himself hints at this when he quotes the distinguished 19th century ornithologist William Yarrell. Yarrell, who authored *Notes & queries* and *A History of British*

---

1. Henderson, William; *Folk Lore of the Northern Counties of England and the Borders*, (W. Satchel, Peyton, & Co, 1866).

Cragside - according to legend the Gabriel Hounds have sometimes flew through the skies at this picturesque beauty spot.

The site where one witness, Graham Randle, had an encounter with a flock of Gabriel Hounds.

Allegedly a photograph of a Black Dog taken near Morpeth in Northumberland. The picture was taken at dusk from a moving vehicle, and although a dark shape can be seen apparently clambering over the wall the photograph is of too poor a quality to be put forward as evidence. The author has some doubts about the authenticity of this image.

A Gabriel Hound – or at least, one version of the legend (artistic representation).

Front Street, East Boldon. It was near here that a Black Dog menaced numerous villagers for several days in the year 1851.

HAUNTED HOUSE AT EAST BOLDON.—During the past fortnight there has been considerable excitement in East Boldon, owing to the house of a gentleman residing there having been visited by a ghost. At night strange rumbling noises have been heard, several of the windows were broken. All the family watched, and as soon as they heard the crash of broken glass, they rushed out at the door, but to their great surprise, nothing was to be seen. An old labouring man agreed to mount guard with a blunderbuss, and shoot the ghost. While he was walking to and fro in the garden at midnight, and all was silent, the spectre appeared to him in the shape of a black and white dog, which ran between his legs; he was suddenly thrown to the ground, and the blunderbuss, though the trigger was pulled by no mortal hand, went off with a loud noise, which so alarmed the sentinel, that he screamed out "murder" but when the inmates came forth, the apparition had vanished. The next night three stalwart yeomen agreed to watch. They accordingly mounted guard with their old rusty firelocks loaded to the muzzle, and each having a large stable lantern, and besides being well furnished with rum punch. At the witching hour of night, they were alarmed by the growl of an invisible wild beast, trying to get loose. They were struck with terror at the unwonted sound, and fled, leaving their arms behind them. The house still continued to be haunted and about 28 squares of glass were broken. At last three of the county police came and watched by turns several days and nights. P. C. Smith had been watching one whole day, and the moment he descended the staircase, more windows were broken. On Thursday last he went into an out-house, and lifted a tile off the roof. He thus commanded a complete view of the premises, and after watching for upwards of 19 hours, about 4 o'clock in the afternoon, after the shutters had been closed, he heard the crash of broken glass, which fell on the outside, and a few moments afterwards the servant girl stepped slyly out, and picking up the broken glass, ran into the house. A few minutes after the young lady of the house, who is about 17 years of age, came out and began to dance the polka, in a graceful style, when suddenly she snatched up a stone and threw it through the window breaking another pane. After which she ran into the house, screaming "mother, the ghost is come again." But the policeman who followed her in, coolly told the young lady and the servant girl, that he had discovered the whole mystery, and that a warrant should be issued to bring both the ghosts before the magis-

The story of the Black Dog of Boldon and the "ghost" as it appeared in the *North & South Shields Gazette,* Friday February 14, 1851.

Grange Park in East Boldon; site of a blacg dog incident in the 1990s.
BELOW: Chillingham Castle, where several black dogs have been sighted.

Tynemouth Castle: A black dog was spotted here in 1980.

Birds, was no mean pontificator. He was even credited by Charles Darwin for discovering a new wild swan in England in May 1830.

According to Yarrell, the creatures were not hounds, but simply bean geese travelling south in huge flocks as winter approached, partly from Scotland and some from Scandinavia. Henderson notes that the birds seemed to prefer the cover of darkness for their migration, and uttered a peculiar cry as they traversed the skies.

All of this suggests that the Gabriel hounds and the ghost birds could have both a common origin and a natural explanation – except for one, singular fact. Bean geese fly south at the onset of winter, whereas the Gabriel hounds and the ghost birds are mostly seen during the summer.

The jury is still out, methinks, on the existence of the Gabriel hounds.

One of the most peculiar black dog stories has its origin in South Tyneside – or North Tyneside, depending on what version you believe. This tale blends wonderfully a tragic death, a ghost and a black dog into one, wonderful legend. At the beginning of Queen Victoria's reign, South Shields was a vibrant town with lots of character. That being said, however, there were still several areas where those who were street-wise would not venture after dark. One such locale was Melbourne Place, which in 1830 was inhabited by a strange mixture of decent folk and undesirables. Burglary was quite common then, and the practice of putting a sign in the window which said, "*Nothing here worth nicking"* still did not deter the dogged thief.

The following true tale concerns a family who lived in Melbourne Place, although I have been unable to discover their names. Thanks to stalwart investigator Alan Tedder of Sunderland, however, for supplying many of the other details. The family in question consisted of a husband and wife, plus one son. The son was a seafarer who had been away sailing in foreign climes. One trip finished early, and the young sailor - nicknamed "Fatty", due to his large girth - arrived back in South Shields sooner than expected.

Fatty's ship docked late at night, and he was looking forward to seeing his parents, whom he had sorely missed after so many months away at sea. He could have stayed on board till the following morning, but decided instead to make straight for home. By the time Fatty reached Melbourne Place his parents' house was all in darkness, his mother and father having long since retired to bed. Quietly he turned his key in the lock and opened the door. At some point he must have made a noise - perhaps by knocking something over - and the racket was loud enough to wake his parents upstairs. Both his mother and father, unaware that Fatty had returned home, naturally assumed that they had an intruder. Quietly they tiptoed down the stairs to confront the burglar. Unfortunately for Fatty he had not yet turned on the lamp. Consequently his father could only see a large silhouette in front of him and lashed out. He hit Fatty with something - perhaps a walking cane, we cannot be sure - and the young man fell to the floor mortally wounded. Still unaware that it was their son, both the householder and his wife continued to pummel "the intruder" with heavy blows.

At some point the lady of the house lit the table lamp and the couple were horrified to find that the supposed burglar was none other than their own child. A doctor was called and frantic efforts were made to revive the youth, but to no avail.

No charges were brought against Fatty's parents, as it was unanimously agreed that his death had been nothing but a tragic accident. However, this was certainly not the end of the matter, for not long afterwards Fatty's ghost began to appear in the house. His apparition was seen by both of his parents, and on several occasions it was accompanied by the spectre of a large, Black Dog. Long after Fatty's ghost had faded into the ether permanently, the menacing hound continued to haunt the area.

There are several variations to this tale still in circulation. One version places the tragic event not in South Shields but North Shields, whilst another relates that the dog was a pet of Fatty's when he was younger. Yet another states that the young sailor had befriended the dog in some foreign port, sneaked it on board the ship and kept it in a secret compartment in the hold. After Fatty died the poor animal was supposed to have starved to death, still trapped on the ship. I find this latter version of the tale unlikely, for why would Fatty have left the vessel without his beloved hound?

In the version of the story told north of the Tyne, the dog is said to still haunt the vicinity of Smith's Dockyards, and bears all the unholy characteristics of the typical black dog. Indeed, he is sometimes referred to as "Shuck" or "Shock" by locals.

The origin of this black dog is most strange, for it seems to be the ghost of a real-life canine (in some versions of the story it is said to be a black Newfoundland). However, after death it seems to have taken on the attributes of more "conventional" black dogs found in the wilds of Northumberland; large glowing eyes, etc. Fatty's death is said to have occurred on 31st October, and it is on this date, each year, when the creature is said to haunt either the site of Melbourne Place and/or the docks at North Shields.

Later in this volume I'll detail the hunt for a mystery wildcat that was dubbed the 'Cleadon Panther' and the 'Cleadon Puma' amongst a number of other epithets. At one point during the investigation the possibility arose that the creature may not have been a cat, but rather an extremely large dog. I think this is unlikely, but there is, I believe, a slim possibility that there may have been a "black dog" connection in this case. This is strengthened by the fact, albeit slightly, that another black dog incident took place well over a century before just over a mile away in the village of East Boldon.

In February 1851, the Tyneside village of East Boldon (then in County Durham before the borders were redrawn) was gripped by an affair that sent shock waves throughout this essentially rural community. Today the case is virtually unknown, and I doubt that so much as one current resident there has heard of it. As I live in the adjoining village of West Boldon I was in an excellent position to investigate the case, which was first drawn to my attention by Sunderland researcher Alan Tedder.

According to the few details that have survived, the epicentre of the affair was "a house in East Boldon" although I have been unable to determine exactly where the dwelling was located. In 1851 the combined population of West Boldon and East Boldon was 1,008. It is likely that the house concerned was one of the terraced houses clustered near to what is now Newcastle Road, although we can't be sure. However, although the exact location of the premises cannot be determined I have been able to find out something about the family concerned.

We know that the house contained four individuals; a "master of the house", his wife, one daughter and a servant girl whose age was given in reports as seventeen.

I did quite a bit of extensive research at South Shields Local History Library and eventually managed to identify the family concerned with a fair degree of certainty. In February 1851, there were three families living in "the Boldons" who had living with them a servant girl aged 17. Two families can immediately be ruled out as there were numerous other people living in the dwelling at the time, and if they had been present as the case developed their absence from the narrative would have been truly extraordinary. This leaves us just one family mentioned in the 1851 census, which fits the bill. Unfortunately, the exact location and address of dwellings are not always listed in the 1851 census. Dozens of people are simply listed as living in "East Boldon", and given a "schedule number". After East Boldon appears at the top of the page, the word "ditto" simply appears afterwards with great monotony instead of a specific address. Unfortunately this is the case with our family, but we should at least be thankful that the census gives us their names, ages and other personal details.

The "head of the house" is listed as one Robert Rukaby, aged 48 years. His wife is listed as Ann Walton Rukaby, aged 50 years. Two children are listed; a Robert Rukaby Jr, aged 20, and an Anne Rukaby aged 15. A servant girl is also registered as living at the premises; one Anne Rukaby, aged 17.

Robert Rukaby was a commercial traveller who had been born in Bishopwearmouth just after the turn of the 19th century. Ann Rukaby, *nee* Walton, had been born in Stockton upon Tees two years before. It seems that around 1831 or before the two met and married. The couple must have resided in Bishopwearmouth before moving to Boldon, as both their children were registered as being born there.

The curious factor about the census entry is that the servant girl is also surnamed Rukaby. This may simply be a mistake on the census entry, or it could indicate that she was a relative of Robert Rukaby, perhaps a niece. Confusingly, this meant that there were two Anne Rukabys and one Ann Rukaby all living at the same address!

During the following narrative, drawn from contemporary records, no direct mention is made of either Robert Rukaby Sr. or Robert Rukaby Jr. being at home. This is understandable, as the father's work as a "commercial traveller" may have demanded his absence for considerable periods of time. Robert Jr. was registered on the census as a sailor, so he too was probably at sea when the incident occurred.

The mystery animals of Northumberland and Tyneside

On the evening of Saturday, 1 February, certain unidentified members of the household heard a series of strange "rumbling noises" at the rear of the property. They investigated, but found nothing untoward. Shortly after, a window was mysteriously broken at the rear of the house; then a second. Everyone present went out to investigate, but could see no one. They then decided to keep watch surreptitiously. It wasn't long before a third window was broken. Immediately, the family members ran outside but were amazed to find that there was no one in the vicinity. They were puzzled, as there was literally nowhere that the mystery assailant could have hidden. At first the damage was blamed upon "drunkards", although this was really just a deduction borne on the wings of logic, and not based on any factual evidence. Two nights later a second window was broken, and the following night a third, all of which led locals to conclude that the motivation for the attacks was not merely the desire to commit a random act of violence. The house, it seemed, was being deliberately targeted in a vendetta.

The following evening, the local residents decided that enough was enough. A hastily organised vigilante group determined that they would catch whoever was responsible.

One, unnamed, elderly local – described as "an old labouring man" armed himself with "a blunderbuss" and stood guard over the property. His primary motive seems not to have been to catch the culprit, as he "walked to and fro" in the garden of the property at midnight, and in full view. Rather, his presence seems to have been designed to act as a deterrent.

Not long after he took up his position and the residents had retired to bed, the old man had an unnerving experience. Suddenly, a "huge dog" jumped out of the shadows and "dashed between his legs". If the creature was as large as the report from the time describes, then we must not wonder that he was apparently knocked to the ground with considerable force. Unnervingly, the blunderbuss was knocked from his hands too and clattered to the ground. It was then that the old man had his second strange encounter of the evening, for, as the gun lay out of his reach, he watched in amazement as the trigger was slowly pulled as if by an unseen hand and the weapon discharged with a thunderous bang.

This experience certainly terrified the witness. He ran through the village shouting, "*Murder! Murder!*" in an effort to wake up his neighbours and companions. There had been no murder, of course, but the ruckus he caused did the trick. Within minutes he was surrounded by a curious group of onlookers wondering what had happened. According to the pensioner, the dog had been "black and white in colour" (not unknown in black dog stories). Villagers noticed that the man was shaking from head to foot and that his own teeth were chattering.

A hasty meeting was held, and the villagers decided to mount a "Second Watch" upon the property the following evening. This time, though, there would be a number of citizens on guard, not just one. If the creature turned up again, they'd be ready for it. Some villagers, of fainter heart, suggested that the hound may be of "Devilish origin" and was best left alone. Genuinely terrified, they retreated indoors and left their more robust neighbours to keep watch. These were described in the press as "stalwart yeomen", not given to fear or nerves. The three men armed themselves with "firelocks loaded to the muzzle" and each carried a large stable lantern. They also were supplied by the household with copious amounts of rum punch. They did not have to wait long before the beast returned. Suddenly, again at midnight, one of the watchers saw the "enormous" dog glowering at them from some nearby shrubbery. They raced towards it with one aim; to shoot it dead before it could do any harm. Here, though, this already bizarre story takes an even stranger twist.

As the vigilantes descended upon the dog it withdrew into the bushes. Emboldened, they raced forwards even faster. To their horror, however, they were met not by the ferocious canine but by a large, growling bear! Curiously, though, reports from the time stated that the bear was "invisible". One can only presume that it was the noise the creature made that helped them to identify it. However, Alan Tedder's account of the affair does not say that the bear was "invisible", but merely "ghostly" in appearance. Personally I think Tedder's account may be correct here.

Bears, of course, had – at least officially – been extinct in England and the rest of the UK since Mediaeval times. It is likely that a few small populations may have survived in remote areas for some years, but not until the 19th Century – and certainly not in or near East Boldon. The appearance of the bear, no matter how anomalous, actually strengthens the possibility that this was a genuine black dog event. Historically, black dogs are widely believed to shape-shift into other animal forms such as bears, horses, cows and even humans.

Another enigma presents itself regarding the testimony of the three men later. Each claimed that the "invisible bear" they were confronted with was "trying to get loose". Get loose from what? How indeed was the creature being restrained, and how could they know if it was invisible? The contemporary accounts of the incident do not say.

Genuinely frightened, the posse ran back towards the residence they were supposed to be guarding, leaving their weapons and lanterns behind them. As they departed, they heard the creature bellow a low, throaty groan. By now, the sleeping villagers had woken up again because of the commotion and assembled outside to determine what had transpired. The vigilantes lost no time in telling them. One described the bear's face as having a "ghostly" look to it, which alarmed everyone even more. This presents us with another mystery, for the creature was supposedly invisible, as I've already stated.

That the vigilantes were truly terrified is evinced by the fact than not a single one could be persuaded to maintain watch a moment longer. Whatever the creature was, it was not of the Good Lord's handiwork and they wanted neither contact nor confrontation with it.

The following day, an incredible 28 small panes of glass in the residence were shattered, and no matter what precautions were taken no one was apprehended at the scene. Baffled and frustrated, someone at the house – probably Mrs. Rukaby - immediately contacted the County Police and told them the story. That evening, three officers, led by one "PC Smith" were sent to investigate. For that night and all the following day, the officers secreted themselves in turn around the outside of the property without let-up. The window smashing continued unabated, but the constables never so much as saw the person or persons responsible, let alone apprehended them.

On the following day, Thursday, two of the officers were seconded to other duties, but PC Smith remained at the property. It seems that catching the felon became almost an obsession with him, and who can blame him? However, he realised that a change of tactics was called for. PC Smith placed himself in a bedroom at the rear of the property where he had a good view of the back of the premises. At some point he decided to go downstairs – perhaps to get some tea from the kitchen, who knows. This involved passing the door which led from the house directly outside to the rear of the property. Just as he did so, there were several thunderous crashes and the officer immediately realised that a number of windows had again been broken. Within seconds he had flung the door open and dashed outside. Again, to his bafflement, he found the place perfectly deserted.

That evening, PC Smith came up with another idea. At the rear of the property there was a storage building, which was actually attached to the main house at a 90° angle. At precisely 9pm, the officer gained entry to the outhouse and placed a set of stepladders directly under the point where the roof adjoined the wall. He climbed up and placed his hand directly under one of the terracotta tiles. Gently but firmly he began to manipulate the tile back and forth until it loosened. By altering its position, and craning his neck, he could look through the gap that had now appeared. Further manipulation allowed him to adjust the angle of vision so that when he looked through the hole he had a perfect view of the rear of the main building. Now, all he had to do was wait.

To indicate how determined PC Smith was to catch the miscreant responsible we need only consider one, salient fact. The constable stayed in that outhouse without once leaving for an incredible *nineteen hours*. He must have had some stamina. PC Smith did not sleep. He sat and listened, and the slightest noise or disturbance found him dashing up the ladder and peering out from his spy-hole in the roof.

At 4pm the following afternoon, PC Smith was still at his post. The shutters over the house windows – or at least most of them – had been closed. However, the officer was suddenly startled to hear the now familiar crashing noise as a window exploded. He ran up the ladder and looked out. At first, all he could see was a pile of broken glass on the pathway underneath the broken pane. Then, within seconds, he heard the click of the backdoor latch being lifted. To his astonishment – and one supposes his immense satisfaction – he saw the young servant girl "slyly" creep outside and proceed to pick up the shards of glass from the ground before promptly running back indoors. Fighting back the urge to announce himself, PC Smith simply continued to watch. Several minutes passed, and then the most extraordinary thing happened. Again the back door opened, but this time the house was exited not by the servant, but by the "daughter of the house" Anne Rukaby, who was 15 years of age. What the officer concluded about what was to follow we can only guess at.

The young woman stepped onto the path and, without delay, immediately begun to "dance the polka". This at least must have impressed Smith, who later confessed that she had done so "in the most graceful style". But then she spoiled things by picking up a rock from the garden and throwing it through a windowpane with all the strength she could muster. Then, playing to the gallery, Anne Rukaby ran inside the house hysterically screaming, "Mother! The ghost is come again!" (It is interesting that young Anne called to her mother, and not her father, or both. This also may be indirect evidence that Robert Rukaby Sr. was not at home when the affair took place.) By this time Officer Smith had had enough, and obviously concluded that he now possessed all the evidence he needed to solve the case. He left the outhouse for the first time that day and promptly followed the daughter into the dwelling.

Seemingly, the game was up. A warrant was issued for the arrest of Anne Rukaby the servant girl and Anne Rukaby the daughter, although why Smith didn't simply lock them up there and then is a puzzle. On Wednesday 12th February they appeared before the magistrates at South Shields, but all the charges were dropped. This in itself is slightly anomalous, and we need to ask ourselves why. There were, in 1851, a number of offences that they could have had stuck upon them, including, of course, the damage to over thirty panes of glass. However, the magistrate Archbold Jervis "and all the law authorities" ruled that their was "no case produced to rule a ghost", and the women were discharged. It seems likely that someone had decided to give the two young women a break. It could be that Mr or Mrs Rukaby – the fact that they employed one servant at least indicates that they must have been persons of some means – exercised a degree of influence over Archbold Jervis and convinced the magistrates' bench to reduce the "charges" to crimes that simply didn't exist, thereby guaranteeing that they would be exonerated. Alan Tedder's account of the incident, the only one I've been able to find other than those which appeared in the *North & South Shields Gazette*, states that "all charges were then dropped, for none related to any on the statute books, which did not mention any offence for imitating a ghost."

Of course, the Rukabys could have simply tried to have all the meaningful charges foisted upon the servant girl, thereby leaving their daughter in the clear. However, as the daughter of the house was the only one whom the fastidiously honest PC Smith had *seen* actually breaking a window this would have been difficult. Further, there is nothing to suggest that the Rukabys were any less honest, and the thought of making their servant girl take the rap for crimes that their daughter was at least 50% responsible for may have been morally repugnant to them also.

But there are other puzzles surrounding this strange affair, some of which will take us back to the enigma of the Black Dog. Firstly, we need to ask whether the girls were responsible for breaking all the windows in the property. This seems unlikely indeed. Whoever or whatever was responsible for the damage, their work rate was prodigious. In just one day, nearly thirty panes of glass were smashed, remember. Is it really feasible to imagine that two young teenagers could have caused this much damage without once being caught? Even if they had managed to avoid detection, how long would it have taken Mrs Rukaby, the police and the vigilantes to realise that whenever the damage occurred the girls were a) always on the scene, but b) never in a position to be observed? Suspicion should inevitably have fallen upon the two young women within a very short space of time, but it didn't. This forces me to conclude that there must have been occasions when, ostensibly at least, it seemed to be impossible that the girls could have been responsible. This then leads us to conclude that, at least some of the time, they may *not* have been responsible for the damage at all. So what are we to think?

My belief is that, at least in the early stages of the affair, neither the servant girl nor the daughter was responsible for breaking the windows. What happened, I think, was that both girls came to enjoy the attention that the affair was bringing upon the household, and didn't want it to stop. Perhaps they then decided that they would become "partners" with the mystery attacker in an effort to ensure that the breakages – and thereby the mystery – continued even if the original miscreant eventually desisted from his criminal acts and went on to pastures new.

Another mystery revolves around the actions of the girls when they were eventually "caught in the act" by PC Smith. According to the *North & South Shields Gazette*, the first window exploded, and only then - *after* Smith began his observations through the spy-hole in the roof - did the servant girl leave the house, pick up the shards of glass, and hurriedly take them inside. Of course, even in those relatively unsophisticated times, one didn't have to be a rocket scientist to realise that broken glass from a window will follow the trajectory of the object that shatters it. If the stone is thrown from the outside into the dwelling, then the broken glass will be found inside. Conversely, if the stone is thrown from the inside of the dwelling out-over, the glass will be found outside the premises. It would seem, on the surface of things, that the servant girl obviously knew this, and did not want to leave any forensic evidence around outside which would indicate that the window had been broken from *inside* the house.

According to the account supplied by Alan Tedder, the servant girl had actually crept into the yard *before* the window was broken and picked up a large stone from the ground before re-entering the property. This clearly implies that the servant girl then either broke the window or at least handed the stone to the daughter of the house who then broke it herself. The problem with this theory is that the only witness to the incident was PC Smith, and in his account, related in the newspaper, he *makes no mention* of seeing anyone before the window was broken. In fact, the first person to be seen by Smith was the daughter Anne Rukaby herself. This may shed a different light on the motivation and actions of the servant. She may have had no part in breaking the window at all, and may only have picked up the glass and taken it inside to divert attention away from the daughter, with whom she seemed to be quite friendly. In short, she may simply have been covering the daughter's back. This is not surprising, as they may have been cousins or at least related in some other way.

The account which Tedder has in his possession also paints a different story about subsequent events, and makes no mention of the daughter of the house smashing a second window or "dancing the polka" just before throwing a stone. In fact, according to the account in the *North & South Shields Gazette*, there was no direct evidence that the servant girl had done anything other than pick up the glass resulting from the first broken window. This may also explain why she, along with the daughter of the house, was never charged with any criminal offence. There is no doubt that the daughter of the house was implicated in breaking at least one of the windows. As to whether the servant girl was really implicated or not we may never know.

During the time these events were transpiring we do not know what happened inside the house, but we can make some quite reasonable assumptions. First, we may safely assume that the sound of the windows breaking alerted the rest of the household, which effectively consisted of Mrs Rukaby. It may also have woken a good few neighbours. As Mrs. Rukaby made her way to the kitchen to investigate, the servant girl and the daughter would have been busy depositing the pieces of glass collected outside across the kitchen floor. By the time the mother arrived, the hastily rearranged evidence would indeed have suggested that a stone had been thrown through the window from outside. Still, there does seem to be an unusually long hiatus between the first window being broken and Ann Rukaby entering the room to investigate. What this means we cannot say with any certainty. Maybe she was in another part of the dwelling away from the kitchen, and either did not hear the first window break or was otherwise delayed in getting to the scene of the crime. The actions of the daughter were undoubtedly suspicious. Why had she run *outside* after the first window was broken before re-entering and shouting for her mother? Why hadn't she been afraid to go outside, considering the possibility that the "ghost" may still have been out there? Why hadn't she done the obvious thing and simply ran through the house to where her mother was, and alerted her? The answer, of course, is that she *knew* there was no "ghost" outside as she had broken the window herself.

It seems likely that there was indeed some form of paranormal activity taking place at the house. To think that so many windows could have been broken in the presence of so many people without the culprits being detected just doesn't make sense. This in itself is interesting, because attendant poltergeist-like activity has often been reported when Black dogs have been present. We should also not forget that the household was disturbed by "strange rumbling sounds" around the property just before the first window was broken. This is another classic symptom of poltergeist presence. All in all, the presence of genuinely paranormal activity makes it more likely that the stories about the black dog and the "invisible bear" may have some truth in them.

What were these creatures? Did the vigilantes merely spot them because they happened to be outdoors at an hour when they would normally have been tucked up in bed? What connection was there, if any, between their appearance and the damage done to the house nearby? There are no easy answers, but it is hard to believe that all the vigilantes simply imagined everything. What could have precipitated the appearance of the black dog? Well, the Boldons and the nearby village of Cleadon have a bit of a history when it comes to cryptozoological animals as you'll see further in this book. Whilst investigating the mysterious case of the 'Cleadon Panther' – more of which later - rumours began to circulate that witnesses may have actually seen a "giant dog" and not a wildcat. This idea gained some coverage in the press, and it is just possible that I may have seen it during my field research, although I can't be sure. In addition to the above, the servant Ann Rukaby had been born in Wallsend, then in the County of Northumberland; the heart of black dog stories in the north east of England. Whether she had told tales to the locals about black dogs, and thereby prepared them psychically and psychologically to see such a thing, I cannot tell. However, several years ago another black dog was seen in East Boldon, and the description given by witnesses was eerily similar to that of the dog in the Rukaby case.

On Monday 5 November, 1999, a local businessman from Cleadon Village had a bitter row with his wife after he discovered that she'd been having an affair with a colleague of his. In an absolute rage, he downed several large whiskeys and contemplated what to do next. He decided to confront his so-called friend, who lived in a street adjacent to the main Newcastle-Sunderland Road that runs right through the Boldons. Having had a considerable bit to drink, he didn't want to chance driving, and so he opted to walk from Cleadon to East Boldon. He estimated that at a brisk pace the journey would have taken him about 20 – 30 minutes. When he told me the story some weeks later, he candidly admitted that he had planned to do his wife's lover "some real damage". Unfortunately – or perhaps fortunately – his colleague wasn't in when he arrived at his home. Frustrated, he decided to walk back home and talk things out with his wife. Normally this would have involved turning right onto the Newcastle-Sunderland Road and then wheeling left at a junction known as Black's Corner. However, there is a park known as Grange Park which can be used as a short-cut. By turning left into Grange Park one can effectively cut off a corner and reduce one's walking time by maybe a minute or two. There is a long path, which leads through Grange Park, and it isn't terribly well lit.

Our witness had reached somewhere around the midway point on this path when he heard what he later described as a "strange flapping noise". Puzzled, he stopped walking and continued to listen. After a few seconds the flapping noise – "like a loose canvas sail or a flag whipping in the wind" – stopped, only to be replaced by the piteous sound of a woman wailing as if she was in great distress. Now foxes can often be seen wandering around both East and West Boldon at night, and they can produce an unearthly wail very similar to that our witness described. In fact, he did indeed think that he had stumbled across a fox, but that notion was rudely dispelled when, "out of nowhere", a huge dog appeared on the path in front of him.

"I honestly can't tell you where it came from. One second it wasn't there...then it was. I'd never seen anything like it. It was like a small pony. It was black and white in colour and looked like one of those hounds you sometimes see on horror films. It was slobbering and growling, like it was demented.

The main thing I remember about it, though, was its eyes. They were huge and red...and glowing. The thing looked bloody evil, to be honest. I didn't know what to do. I thought about turning around and running back up to the main road, but I had this awful feeling that if I did it may bound after me and push me to the ground. All I could do was stand still and stare at it. Then it just disappeared in a puff of smoke. It was like the thing exploded, but without any noise. I noticed that the air smelt like gunpowder...or maybe fireworks. Anyway, the thing was gone, and I waited a few minutes before running down the rest of the path. I didn't stop till I got home, and I was knackered. I've never been back to Grange Park, and I never will. What I saw scared the shit out of me."

Its hard to resist the conclusion that our witness saw the same thing that presented itself outside the Rukaby household well over a century earlier, as even the colour of the two beasts tallies exactly.

Black dogs are a truly an enigma. I'm conscious that the number of archived tales from Northumberland and Tyneside is far smaller than elsewhere, but they are intriguing nonetheless. I'd encourage those who know of other north-east black dog tales to get in touch with me.

Chapter Four

The Beast of Bolam Lake

I have been investigating paranormal phenomena for nearly forty years, and during that time I have carried out literally hundreds of investigations into anomalous events that have baffled both my fellow researchers and I. Of all these investigations, there are two that I feel deeply privileged to have been involved with. One is the South Shields poltergeist, which fellow researcher Darren W. Ritson and I investigated and the results of that remarkable affair were published in book form as *The South Shields Poltergeist: One Family's Fight Against an Invisible Intruder* in March 2008. The other is the Beast of Bolam Lake. Most investigators would give their eyeteeth – and possibly their right arm – to be involved with one such case of this magnitude. To have been involved with two is nothing short of a blessing.

In the year 2000, I founded an organisation called the *Twilight Worlds Paranormal Research Society* – to my knowledge, the north of England's first, genuine organisation dedicated to the investigation of all Fortean phenomena. Since 1998 I've written a column for the *Shields Gazette*, Britain's oldest provincial newspaper. The *Gazette* has won numerous awards over the years, and I count it a privilege to pen my *WraithScape* column for it. *WraithScape* is a weekly dip into the paranormal; readers send me their tales of ghostly encounters and UFO sightings, and I write them up with gusto. Nearly a decade later, the column is still going strong.

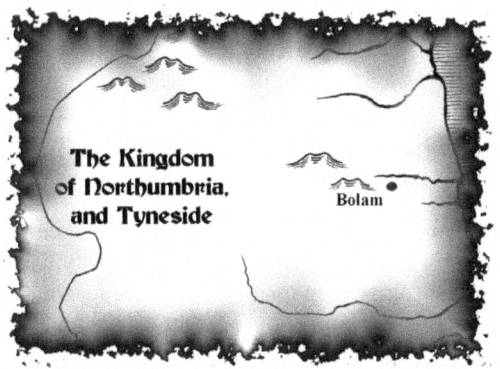

In late 1999, a group of avid readers asked the then editor of the *Gazette* if I'd be interested in holding regular meetings – perhaps in a local pub – where I could relate to them some of the strange tales I've come across during my decades of research. I was happy to do this, and hence *Twilight Worlds* was born.

In 2002, the society was at the height of its influence and it seemed to others and me that there was literally no stopping its growth and maturity. I was wrong, but *Twilight Worlds* was to play a minor but important role in the Bolam Lake investigation.

On Tuesday, December 24, 2002, I received a telephone call from Jon Downes, director of *The Centre for Fortean Zoology*. I've worked with Jon for years, and value him as a trusted friend. He is one of the finest researchers in the field, and I cannot sing his praises too highly. Jon wanted to know whether I'd heard about a story regarding a strange creature that had been sighted in rural Northumberland, specifically the area known as Bolam Lake. I said I hadn't, but I wanted to know more. What transpired over the next few months was a mad, roller-coaster ride through one of the strangest cryptozoological events ever to write itself into British history.

First, though, we need to look at the area of Bolam Lake itself.

Bolam Parish contains the townships of Trewick, Bolam, Bolam Vicarage, Gallow Hill, Belsay, Bradford, Harnham, and Shortflatt. It is bordered on the north by Hartburn, on the west by Kirkwhelpington, on the south by Stamfordham, and on the east by both Meldon and Whalton parishes. The area of Bolam Parish is in excess of 7,000 acres. In 1801, the number of its inhabitants was a mere 434. Now it is much larger, but the area still retains its mystical, almost magical beauty.

Bolam Lake Country Park sits in an idyllic landscape amidst the verdant Northumbrian countryside. It plays host to wetland areas, which in turn provide homes for swans, ducks and other avian creatures. Bolam Lake itself is also a magnet for fishermen. Around the lake is some of the country's finest woodland. Here can be found a healthy population of red squirrels, now beleaguered throughout much of the UK by their grey rivals.

The area around Bolam Lake has been home to *Homo sapiens*, too, for aeons. Visitors can see a reconstruction of a Neolithic house – built on its original foundations – which dates back to at least 3,700BC. Archaeologists have also determined that wheat and barley were farmed nearby during the same period. Walk through the woods at Bolam, and you can sense that the past is merely a whisker away.

Despite its rural tranquillity, the area surrounding Bolam Lake is certainly not a wilderness. Wherever one goes in the vicinity, one is never far away from civilisation. This fact, more than any other, makes the following tale all the stranger, as you will see. Stories of enigmatic, manlike creatures roaming the rural areas of the British mainland are more common than you may imagine.

In the north east of England, near Alnwick, there have been sightings of a creature known as the "Wooly Man of Rugley" for ages. "The Deugar" is a similar man-beast who allegedly lives in a cave and stalks the nearby valley of Coquetdale. According to legend, the Deugar lures unsuspecting witnesses to his lair before slowly roasting them alive over a peat fire. Whatever floats your boat, I suppose. Such accounts go back centuries, in fact. And its not just hominids that have been spotted. In January 2006, a witness reported seeing a huge, pterosaur-like creature soaring through the skies above Nursery Woods in Cumbria. I've personally investigated similar reports further south.

At Callaly, a village in Northumberland, crops were repeatedly ravaged during the Middle Ages by a creature giant known as the Devil Hog.

But back to the man-beasts. In January 1998, an anonymous experient told his own account.

It seems that on one fateful Sunday evening, the witness was walking his dog along a road leading out of Beckermet, near Egremont, towards the A595. As he passed Nursery Woods, he glanced at his watch and noticed that the time was 16.45hrs. As it was winter, the light was fading fast and the man noticed that visibility was becoming increasingly poor. Nevertheless, he was startled to hear the sound of wood snapping, as if someone or something had stood upon a fallen tree branch. Naturally the witness assumed that the noise had been made by an animal, and he stopped to see what it was. Peering through the trees and shrubbery, he was astonished to see a large, manlike creature by a nearby pond. The watering hole, the man added, was approximately 150 metres into the interior of the woods.

Due to the poor lighting, the man had to strain his eyes to get a good look at the beast. As he did so, it slowly turned its head and seemed to look in the witness's direction as if it had seen him or was at least aware of his presence. Later, the man would estimate that the creature was approximately 6 feet in height and probably weighed around 14 stone. More intriguingly, it was covered in a coat of browny/ginger-coloured hair. Almost immediately after the creature saw the witness it reared up on its hind legs and, just like an adult human, walked deeper into the woods and out of sight. The man promptly returned home and reported the sighting to his wife. His first instinct was to tell the police, but she persuaded him not to in case he was made the subject of ridicule. The witness added that he'd lived in Beckermet for seven years and had walked this route almost daily. Nevertheless, he'd never had an encounter like that before. Thankfully, one supposes, he's never had one since. A similar creature had been spotted at Silecroft several years previously.

These encounters did not take place within the geographical boundaries covered by this book, but they weren't that far away. I mention them to demonstrate that what would later be seen at Bolam Lake was certainly not unique.

The Beast of Bolam Lake had first been spotted (as far as we know) *circa* 1997, when an angler - we'll call him George - had been engaging in a spot of nocturnal fishing at the lake with two colleagues. On a number of occasions they had a strange feeling that they were being watched, but dismissed these sensations. At some juncture they decided to call it a day – or more correctly, a night – and they began to make their way back to the car park where they'd left their vehicle. One of the fishermen turned to talk to his friends. There, standing on the wooden boardwalk, he saw a huge, bipedal creature similar to the one seen in Nursery Woods by the aforementioned anonymous witness. It was approximately 8 feet in height and covered in dark hair. George also stated that the creature had, glowing, sparkling eyes. As one would imagine, the three fishermen didn't stop to make further enquiries. They simply bolted for their car and drove off into the night.

"We ran at top speed all the way back to the car. I had a camera in my bag, but the thought had not crossed my mind to use it. To be honest, I think I was too scared."

In the year 2000, George again visited Bolam Lake with his girlfriend[1]. At some juncture George's companion suddenly stiffened and claimed that she could see "a man in a monkey suit" staring at them from behind some bushes. George went to investigate, but could find nothing. Nevertheless, it's almost certain that his first encounter with the Beast must surely have been in the forefront of his mind.

In 2002, George once again returned to Bolam Lake in the company of one of his chums who had been with him on that first, fated fishing expedition. They set up camp and decided to spend the entire evening trying to catch pike. Eventually, George's friend nodded off to sleep. Around 1am, George suddenly heard a crashing noise coming from within the woods followed by two loud thuds. He concluded that they were definitely being made by a very large animal, and that its presence was at one point no more than 10 feet away. Not wanting to encounter the Beast for a third time, George ran – closely followed by his friend. In fact, they didn't even stop to decamp or pick up their fishing equipment.

The next day, George and his friend (or according to some accounts friends and/or relatives) went to the park to retrieve the aforementioned equipment. To their dismay, they found that their metal bait box had been destroyed and that someone (or something) had been rooting through their belongings. There was also evidence that a large creature of some kind had been perusing the vicinity. Just what this evidence was is not stated in reports; perhaps it was broken branches, footprints, etc., who knows.

Later, George's friend admitted to having seen the creature one night the previous March, but *this* time on a hill near a boundary wall close to the remains of an Iron Age settlement. The entity he espied then was large, muscular and covered with dark fur.

In January 2003, the Beast was seen yet again by another witness and her young son. The woman had heard of the previous reports, and hadn't believed them for one minute. Her scepticism had, therefore, emboldened her and enabled her to visit Bolam Lake without any fear of having an encounter with the Beast – a creature that she firmly refused to accept the existence of.

Things changed, however. As the woman and her son crossed the car park they were both star-

---

1. According to one interview he later gave to the CFZ, they were engaged in some alfresco lovemaking when the incident occurred. The link between alfresco sexual activity and bigfoot/skunk ape reports in the United Stateshas been documented widely elsewhere. In *The Owlman and Others* (CFZ, 1997), Jon Downes notes how there appear to be links between the Cornish Owlman reports and girls undergoing puberty. The link between female sex hormones, and zooform phenomena would seem to be a deep and complex one.

tled to see a huge, hairy manlike creature staring at them from its location in the woods. The entity was still to the point of being statuesque, and yet both witnesses testified later that it had instilled within them what Jon Downes would later describe as "an intense feeling of fear and trepidation". Without further hesitation they jumped in the car and sped out of the park. Jon would later interview other witnesses. Some hadn't actually seen the creature, but all reported experiencing the same sense of dread. One witness, a woman in her late 50s, had been visiting the area with her son when she noticed some unusual "tree formations". These will be discussed later, but although the woman didn't actually espy the beast she felt as if she had telepathically received a "message" not to investigate the tree anomalies any further. She too felt compelled by a sense of morbid dread to vacate the area immediately. Other sightings followed in rapid succession. Later that month a group of friends out walking saw the beast staring at them before vanishing into the trees.

By this time, the reports were beginning to reach Jon and his colleagues at the Centre for Fortean Zoology. Curiously, the CFZ was also receiving reports of many other hairy hominid sightings from all across the UK. Later, Jon told me how he'd received a dozen or so reports in the space of one morning from Sussex, Lancashire, Cumbria, Staffordshire, Sherwood Forest and Scotland.

Over the Christmas period Jon had started collecting these stories, and when the other members of the core CFZ faculty returned after their Christmas break, he had a reasonably - sized dossier of sighting reports for them. His original plan had been to suggest a CFZ expedition to one of the "hot spots" of British BHM (Big Hairy Monster) reports sometime in the early spring. However, as by far the most interesting reports were now coming in from Bolam Lake, that site rapidly became their primary target. The Bolam Lake sightings were interesting simply because there were so many of them. Even as the CFZ were putting together plans for a trip to Sussex in March, more and more reports were coming in from the Bolam Lake area and it seemed increasingly obvious that they would have to go and visit the place to investigate them personally.

When Jon told me about the sightings, I was surprised that I hadn't heard about them. Nevertheless, I was anxious to get involved. BHM stories had always held a particular allure for me, and I felt a profound sense of satisfaction at having one take place metaphorically if not literally on my doorstep. I told Jon that I'd be glad to help out in any way I could. Jon suggested that it would be a good idea if I could organise a preliminary reconnaissance trip to Bolam Lake and report back to him. This would help the CFZ prepare their field trip before they set off. I gave this some thought, and eventually decided to take two friends along with my wife, Jackie. Debbie Proudlock and her then partner Trevor Brown were good friends of ours, and on balance I felt that they were the best people to recruit for the exercise. They were both enthusiastic about paranormal research and both leading members of *Twilight Worlds* (although they too subsequently left the organisation shortly after I did). Debbie was a practitioner of Wicca and very "sensitive", and I felt that this may be an advantage in what could prove to be a highly "psychically charged" area. Trevor did not at that time have the same degree of interest in spiritual matters as Debbie, but he was a good organiser and a creative thinker. Jackie, although she consistently denies it, is quite psychic and what is generally known as an "intuit".

The mystery animals of Northumberland and Tyneside

She frequently gets good hunches, and I almost always trust them.

Saturday, 11 January was a bitterly cold day and a thick blanket of snow covered most of Northumberland. Jackie and I left the house early that morning and caught a bus to the East Boldon Metro station. Ten minutes later we boarded a Metro train to Newcastle Central Station where we'd arranged to rendezvous with Trevor and Debbie. When we arrived we toyed with the idea of having a coffee at one of the several kiosks on the station concourse, but decided against it. We wanted to get to Bolam Lake as soon as possible; we had a busy day ahead. Not long after we arrived, Trevor and Debbie turned up and without further hesitation we started off for our destination.

There were several purposes to our visit. Firstly, we wanted to take a good number of digital photographs of Bolam Lake and the surrounding locale so that we could e-mail them to the CFZ. We also wanted to carry out a detailed reconnaissance of the area in an effort to both find and secure any trace evidence of the beast's presence. Thirdly, we wanted to ascertain via psychic/spiritual means whether the 'Beast of Bolam Lake' was a) a taxonomically classifiable animal, b) an animal unknown to science or c) a zooform creature of non-physical or quasi-physical origin.

On arrival at Bolam we made our way into the park and noticed at the entrance that some wag had put up a home-made sign drawing attention to the beast's presence. We laughed, and then parked the car before making our way to the Visitor's Centre near the lake itself. After coffee, cake and scones, we then made an effort to interview both staff and visitors to see if they had had any personal encounters with the creature. To be honest, no one interviewed on-site had anything valuable to contribute directly to the investigation. The warden of the Bolam Lake Country Park and others were clear that they only knew what information could be gleaned from the handful of websites that were already carrying details of the sightings. However, we did manage to ascertain two interesting facts not directly related to the "BHM" investigation. Firstly, there had been numerous reports of other OOPAs [Out of Place Animals] in the vicinity, including wallabies and kangaroos. Secondly, one sighting of an ABC [Alien Big Cat] had apparently been recorded. We made a note of these incidents, and subsequently passed them on to the CFZ.

Our first point of call was a wooden boardwalk which leads through the wooded area around the lake, for we knew that at least two sightings of the beast had occurred here. Perhaps not coincidentally, the boardwalk is directly adjacent to the only wooded area large enough and dense enough to conceivably hide a large, cryptid anthropomorph for any length of time.

As we ambled along the boardwalk, Trevor noticed a peculiar indentation in the soil. Fortunately, this particular location was not covered with snow. The indentation looked like a very large footprint, but erosion, possibly by rain, had made the print of extremely poor quality and of little use as evidence. Nevertheless, we photographed it and continued on our walk. Eventually we arrived back at the clearing, which sits not too far from the Visitors' Centre. This spot is on a higher elevation and not fully protected by trees and other vegetation. Consequently it was covered with a layer of frozen snow approximately two inches in depth. We separated and scrutinised the ground for any anomalies. Jackie was the first person to find any, and they proved to be of exceptional interest.

The first anomaly was a footprint. At first glance it seemed to have been made by a boot, but two things about the indentation were strange. Firstly, it was extremely large – about fourteen inches in length. We considered that a degree of melting may have taken place, and that the print was now much larger that it had first been when whatever had made it had impacted with the snow. However, at the distal end of the impression were several clearly defined marks that, in our opinion, could only have been made by toes. It was later suggested that what looked like toe marks could actually have been made by the heavy serrations on the soles of walking boots; but then again, who wears walking boots *that* large? Again we come full circle and have to consider the possibility that the snow had melted and thereby increased the size of the print, but if this was the case it then becomes hard to explain how the indentations that looked like toe marks were so clearly defined. My candid opinion is that there just may be a natural explanation for this peculiar print, but we can't be sure.

We continued our search, and after a hiatus of maybe two or three minutes I discovered another print. This was larger than the first, and far stranger. It was fourteen inches in length and almost rectangular in shape. The heel tapered away almost to a point, but once again the distal end bore five distinct toe prints. The toe prints were strange indeed, for the second, third and fourth toes were quite small, but the first and fifth toes were disproportionately large; one may even say grotesquely so.

"Jesus", Trevor mumbled as we stood and stared at it.

The third footprint was found by Debbie. It was similar in shape to the second, but smaller and the toes were not as clearly defined.

Three footprints, all different. If the beast had left a trail in the snow, then this implied that there were three of them. Prints 1 and 2 could be adults; print 3 could have been junior tagging along. Personally I found the notion of three creatures roaming around the park too much to cope with. And why were the footprints in isolation? Surely, if the beast had walked across the snow there should have been a *trail* of prints that we could follow. We studied the surrounding area, and it was Jackie who pointed out that an awful lot of people had tramped across this exposed area since who or whatever had left those marks. Whole areas of snow had been trodden on over and over again until hardly a square inch remained intact. Perhaps there *had* been other tracks, but they were now obliterated. A more plausible explanation, of course, was that what we had stumbled across were not tracks at all but simply anomalous marks made by God-knows-what.

My feeling is that Footprints 1 and 2 are likely to be genuine. Footprint 3, however, may simply have been an anomaly.

Trevor, Debbie, Jackie and I then walked to a wire fence, which separates Bolam Lake Country Park from a nearby field. Here we found numerous hairs – singular and in tufts - stuck to the wire. We knew from the beginning that they were almost certainly from horses or cows, although none were in the field at that time, but took some samples for analysis. One long hair matched the colour of the beast as described by witnesses. Satisfying ourselves that nothing further could be gleaned at that location, we then made our way back to the Visitors' Centre for another hot drink.

I have Native American heritage on the paternal side of my family, and take the cultural and spiritual side of Indian life very seriously. Before we even left for Bolam Lake it had crossed my mind that it may be possible to use certain Indian rituals to smoke the beast out of its hiding place.

On leaving the Visitors' Centre near the lakeside we headed west back towards the heavily wooded area. The path sloped away gently downhill to the left and uphill to the right. The first unusual occurrence was that Debbie and I both clearly saw what looked like a large, grey wolf running through the woods in a south-westerly direction. We were both astonished at this and I remember Debbie gasping audibly. As my "Right-Hand Walker" totem animal (an animal spirit guide) is a wolf, I assumed that it wanted us to follow it. Thus, instead of walking to the right we turned left and headed down a narrow path.

I needed to locate a place where the creature had been, to see if I could sense anything regarding its nature or connect with its spirit. We decided to leave the path to our left and walk downhill in the direction the wolf had taken. I told the others to watch out for a directional sign which I believed would be given to us by another one of my totem animals, although I did not at that time know which.

Almost immediately I saw a small, white feather attached to a branch on a nearby sapling. It was stuck firmly to the branch, as if it had been glued there. The tip of the quill was pointing in a south-westerly direction. I showed this feather to the others and told them to watch out for more, as I was now aware of the method of communication and guidance that my second totem animal, a seagull, was using. I also told them that, when they located the feathers, they would find them to be pointing in exactly the same direction. I found no more feathers, but the other members of the team found five between them. They were all glued to branches and pointing perfectly in the same south-westerly direction. Each feather directed us to the next, and so on.

Eventually we arrived at a small clearing near a stream, where the final feather was found. Unlike the others this one pointed directly to the ground, and I could sense that we had found the spot. It had *been* to this place. Using further Native American divination techniques I was able to determine that the creature was currently farther to the north-west, and a considerable distance from us. Again using Native American ritual techniques I invited the creature to join us, and then told the rest of the team to listen for its footsteps.

I was surprised at how quickly it attended. Within thirty seconds or so Trevor and I heard the first footstep. It was faint, but the Winged People (birds), the Wind People (wind) and the River People (the stream) had stopped talking so that we could hear it. Every time we heard its foot hit the ground, the Winged People, the River People and the Wind People would be silent. Then, after we had heard the thud, they would start talking again. Debbie was the first to notice this. *"Listen"*, she said, *"notice how everything else goes deathly quiet when we hear it."*

The footsteps grew louder and louder, and more frequent. I lit some incense and handed it to Debbie, who, because of her Wiccan background, knew exactly what to do with it. She "sent up smoke", and said, "*It's going to go in the same direction*". Again she sent the smoke up and, sure enough, it headed in a north-westerly direction. We left an offering of fruit – a small orange - for the creature, and then, almost simultaneously, felt a surge of energy coming up through the ground into our feet. It travelled to our waists and then dissipated. Debbie felt it first, then I, then Jackie. I am not sure whether Trevor felt this.

At this juncture we heard a strange noise; something that I can only described as a cross between a howl and a warble. It was high-pitched and lasted for around ten seconds. This cry was repeated twice, and on each occasion cattle in a nearby field started to panic, mooing loudly. After another two minutes or so we all agreed that we could sense the close proximity of the creature. When we heard branches snapping under its feet we agreed that it was time to make a discreet exit, and we left.

We concluded that there was no possibility of the 'Beast of Bolam Lake' being a conventional, taxonomically classifiable animal in any normal sense. Although the woods around Bolam Lake could possibly play host to a single creature without being detected for a short period of time, they are just too small to hide even one such creature indefinitely, let alone a breeding population. In my mind I believed there was every likelihood that the creature was of a zooform type, and that there was a strong psychic/spiritual dimension to both its nature and activities. Although the creature did seem to have left physical traces, it was likely that the gleaning of any further useful information or data concerning its nature or motivation would only come about with the help of a number of experienced psychics and ritualists.

When I got home I wrote up a report to forward to Jon and the rest of the CFZ team. In the report, I recommended that because of their far greater experience in this field the CFZ should assume overall command of any further scientific investigation. Because of their close proximity to the site and the expertise of some of its members, I also suggested that the *Twilight Worlds* team should be responsible for logistical support and psychic investigation.

Whilst I was preparing a briefing document for *Twilight Worlds*, Jon was working on a similar document for the CFZ, along with a press release. The media had already caught wind of the story and wanted to know just what the hell was going on at Bolam Lake. As I am the Tyneside Regional Representative for the CFZ, I agreed to handle the media interaction in the north east of England where Bolam Lake is situated.

It was agreed that Jon Downes would be overall mission commander and media liaison. Graham Inglis, Deputy Director of the CFZ, would serve as technical supervisor and deputy mission commander. Richard Freeman, CFZ Zoological Director, would serve as the team's zoologist and head of Fortean investigation. The team also included CFZ Administrator John Fuller, who would serve as assistant to Jon Downes, and Durham Representative David Curtis, who would be assistant to Richard Freeman. Also helping would be Geoff Lincoln from the British Hominid Research Group.

The sheer size of the investigation also required the help of numerous additional personnel. Six volunteers from the *Twilight Worlds* Rapid Response Group were seconded to the command of Graham Inglis. In addition, a further twenty *Twilight Worlds* volunteers were seconded to help in the following areas:

a) Mapping, under the command of Graham Inglis.
b) Data recording, under the command of Jon Downes and Mike Hallowell.
c) Media liaison and general communications, under the command of Mike Hallowell and Jon Downes.
d) Miscellaneous duties.
e) Creation of sand traps, under the command of Richard Freeman.

Jon's briefing document also set out the overall mission objectives:

a) Reconnoitre the Bolam Lake area and assess viability of habitat for supporting unknown species of higher primates.
b) Carry out in-depth interviews with eyewitnesses, with particular attention to the zoological and quasi-Fortean aspects.
c) Carry out brief but thorough surveys of the target area and produce a map.
d) Identify sighting locations.
e) Carry out electromagnetic frequency (EMF) scans of sighting locations and use infra-red cameras.
f) Make a thorough photographic record of the target area.
g) Cross-correlate available data on Bolam Lake sightings with other historical sightings.
h) Seek physical evidence – hair and scat samples (collect), footprints (take plaster-of-Paris samples) and photographs. Sound and video recording.
i) Investigate accounts of occult activity in the region.
j) Cross-correlate the dates of the sightings with dates of other Fortean events in the region or general (political) events on the world stage.
k) Use two sets of clairvoyants/clairaudients (and a third pair, non-clairvoyant, as a control group) to individually survey the event sites.
l) Using raw fish and fruit as bait and the sounds of a known species of great ape, and also sounds purporting to be those of Sasquatch. To set separate sand traps for footprints.

Over the coming days, the newspapers made good use of, Jon, Richie, myself and others.

Jon told reporters, "*Whereas there is very little doubt that creatures such as the Yeti or the Orang Pendek of Sumatra are bona fide species of higher primate that have so far escaped detection by scientists, other man beasts from Australia, North America and Europe are far*

ABOVE: The boardwalk at Bolam Lake where "the Beast" put in one of its first appearances. (courtesy Darren W. Ritson). BELOW: Bolam Lake in Northumberland – this picturesque area played host to a large, mystery hominid in 2002-3

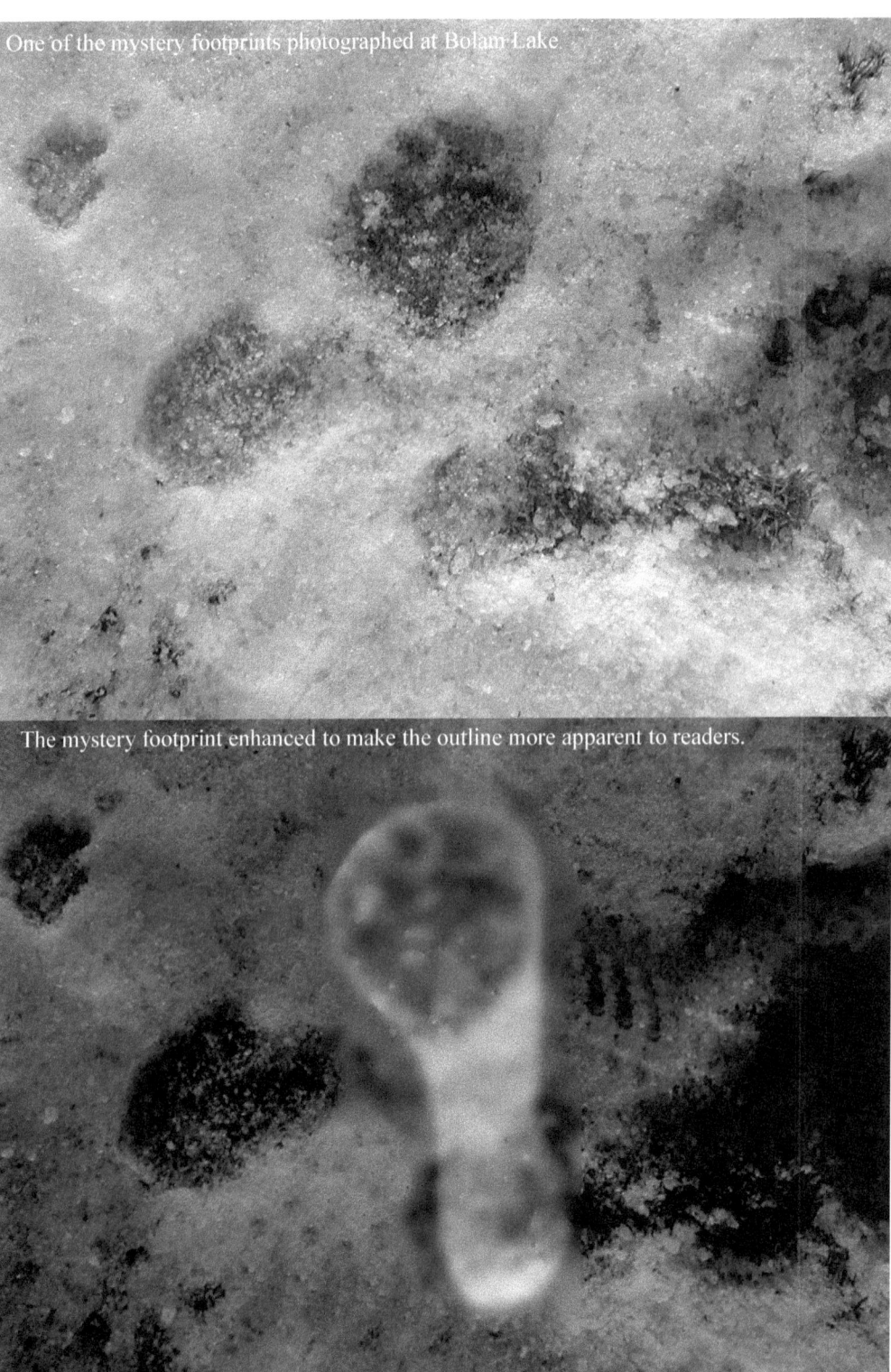

One of the mystery footprints photographed at Bolam Lake

The mystery footprint enhanced to make the outline more apparent to readers.

A second mystery footprint in the snow. (BELOW: the footprint enhanced)

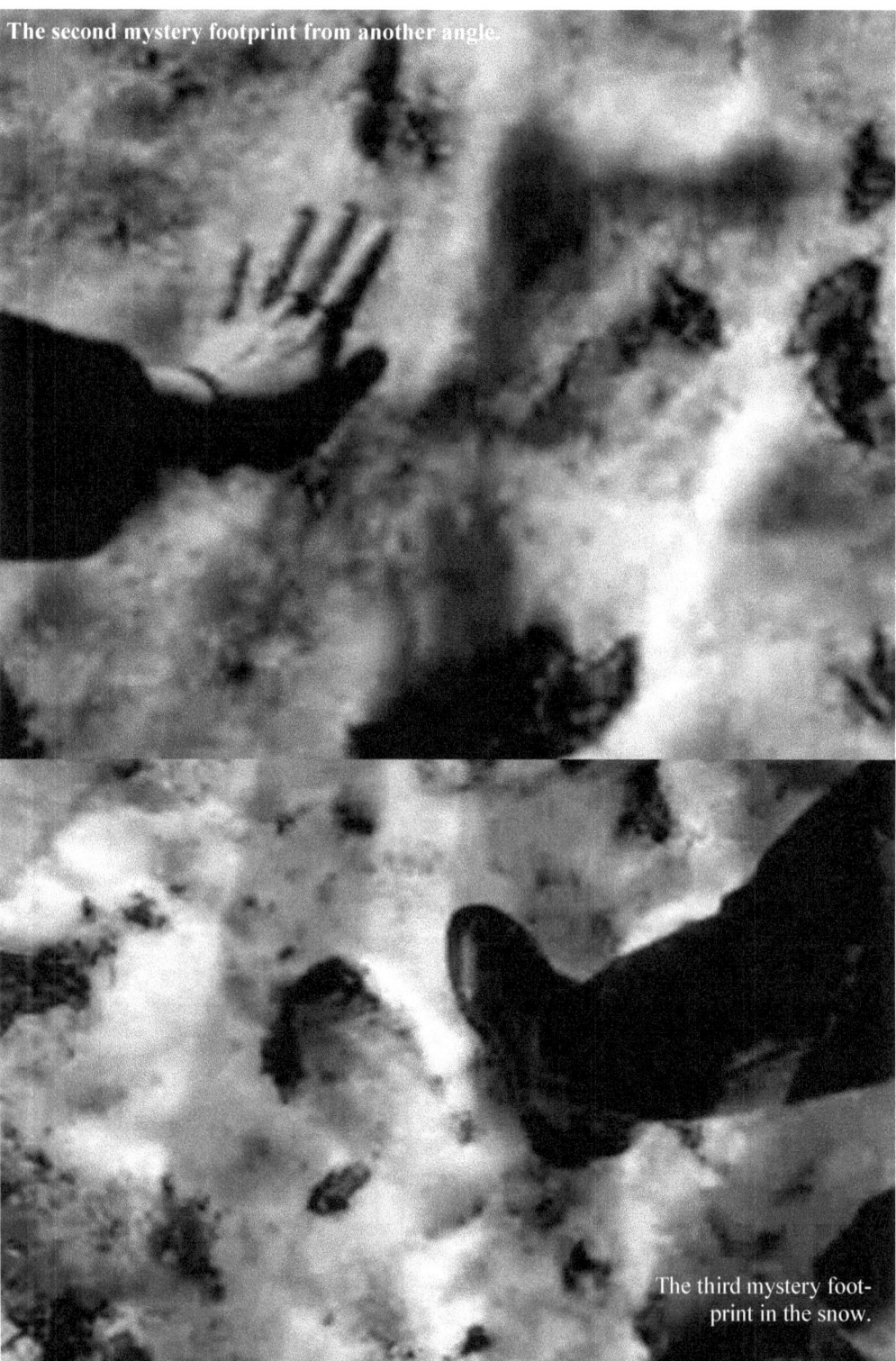

The second mystery footprint from another angle.

The third mystery footprint in the snow.

The "standing stones" at Bolam Lake, which in reality are almost certainly gateposts. Faint traces of indecipherable rune-like inscriptions seemed to have been carved on at least one of the posts, but its difficult to tell.
BELOW: Close-up of the post bearing the strange hieroglyphs.

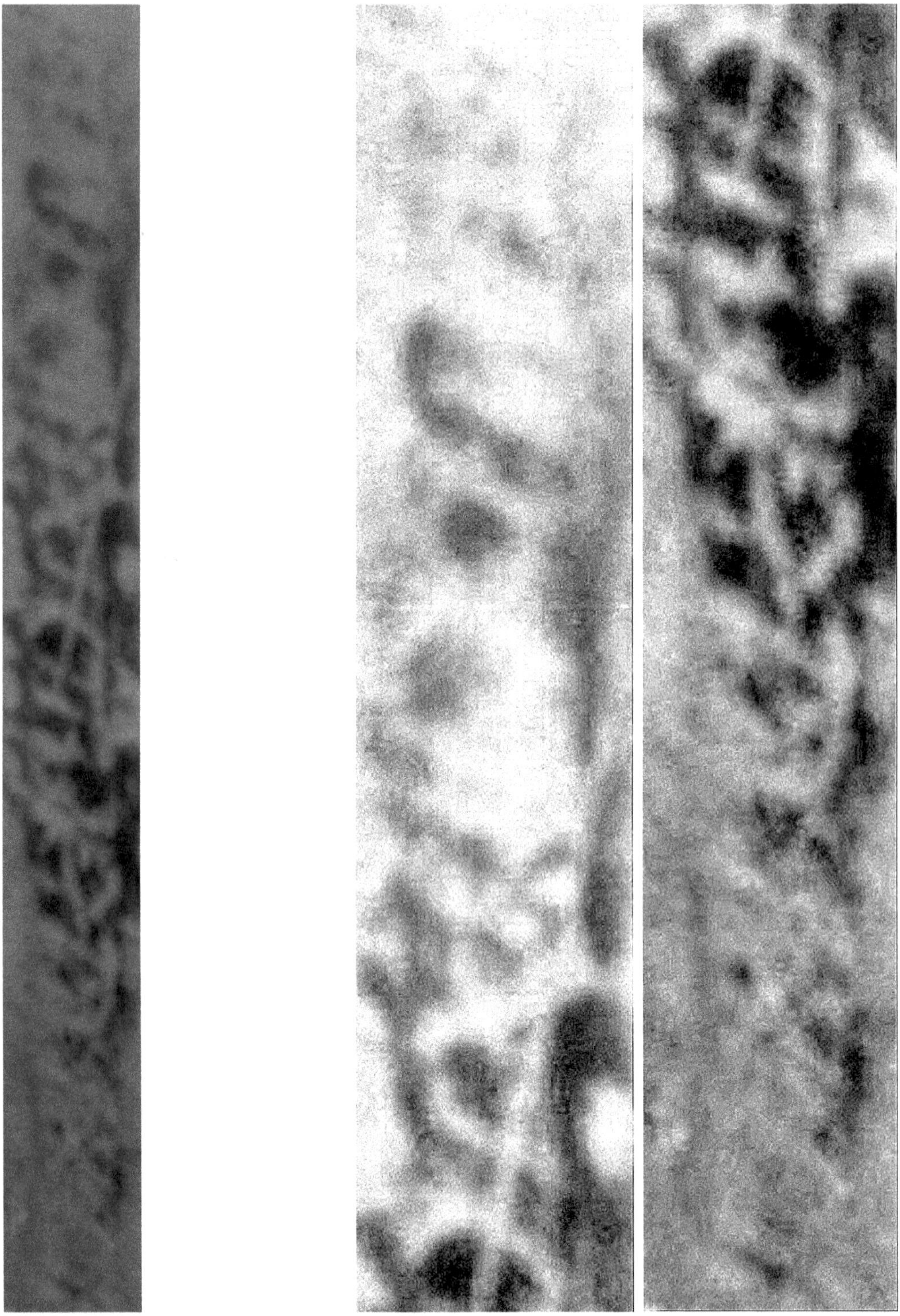

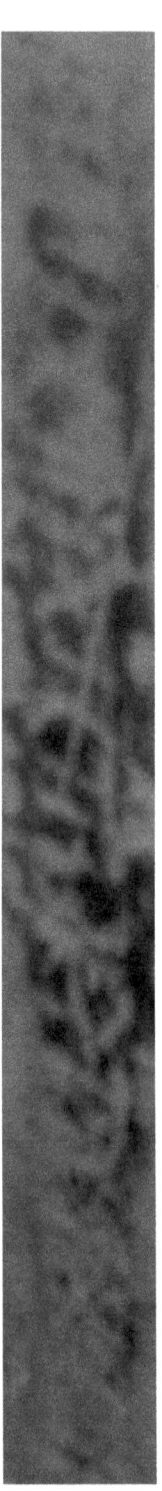

The hieroglyphs.

Incisions on another aspect of one of the posts.

BELOW: Richard Freeman (L) and John Fuller, from the Centre for Fortean Zoology, examine a tree allegedly damaged by the Beast of Bolam Lake. Large amounts of bark seemed to have been stripped from the trunk, but it was impossible to determine the cause.

One team of investigators makes its way into the woods surrounding Bolam Lake.

BELOW: A flattened patch of undergrowth near the place where the Beast was first seen.

The remains of an old "Sasquatch Tipi" photographed in Louisiana by the author.

The remains of a "tipi" found at Bolam Lake, almost identical to the one found in Louisiana by the author. BELOW: The author prepares to carry out a Native American ritual designed to "draw in" the Beast of Bolam Lake.

Prayers are offered after the ritual offering is made.

A sage and sweetgrass smudging stick is lit by the author during the ritual.

BELOW: As darkness falls a few investigators still train their cameras on the woods in the hope of catching a glimpse of the Beast of Bolam.

The woods are illuminated by car headlights a few seconds before the sighting

Investigators wait for their de-briefing at the end of the investigation.

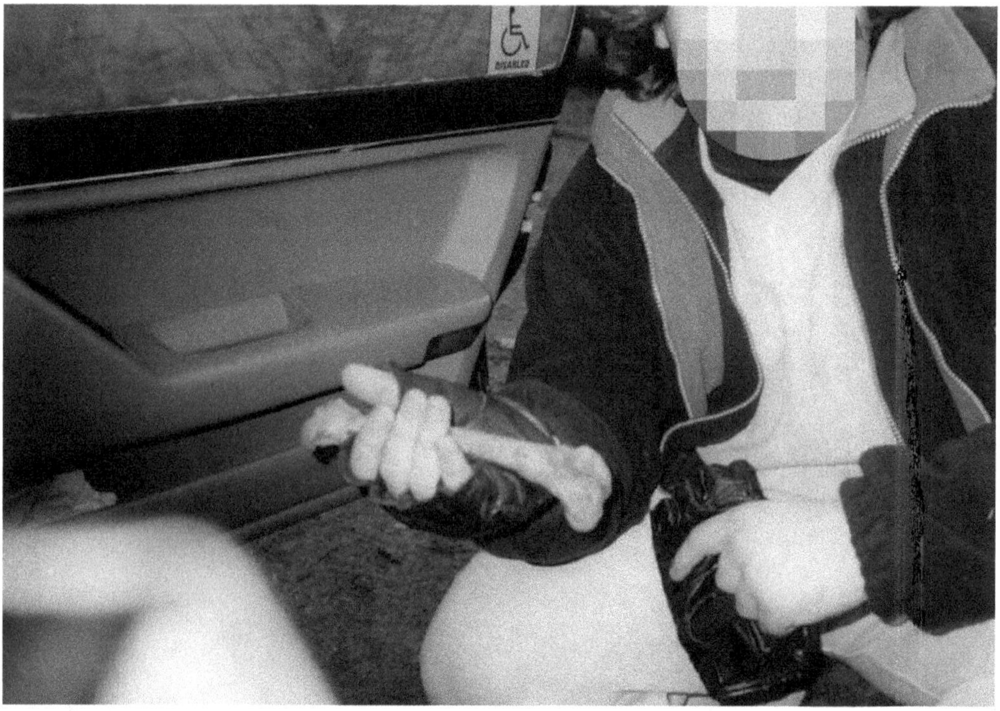

An investigator holds up a bone found in the vicinity where the Beast had been seen several times. Analysis later showed it to be from a lamb carcass. BELOW: Jonathan Downes from the Centre for Fortean Zoology, being interviewed by the press shortly after an encounter with the Beast of Bolam Lake.

At dawn on the day after the "big sighting" several investigators comb the area for tracks or other clues. BELOW: Strangely distorted tree branches and others arranged together to make bizarre structures and patterns. The work of the Beast of Bolam Lake?

More pictures of tree branches and twigs arranged
into bizarre geometric patterns.

Huge branches snapped from trees and then interlocked to form a "tipi".
BELOW: Branches secured to each other at right-angles in the same location.

Several perfectly explainable "orbs" of light which were nothing more than dust particles, and one enigmatic ball of light which left a short vapour trail and was visible to the naked eye. It disappeared approximately two seconds after the photograph was taken at Bolam Lake during the investigation. (BELOW: closeups)

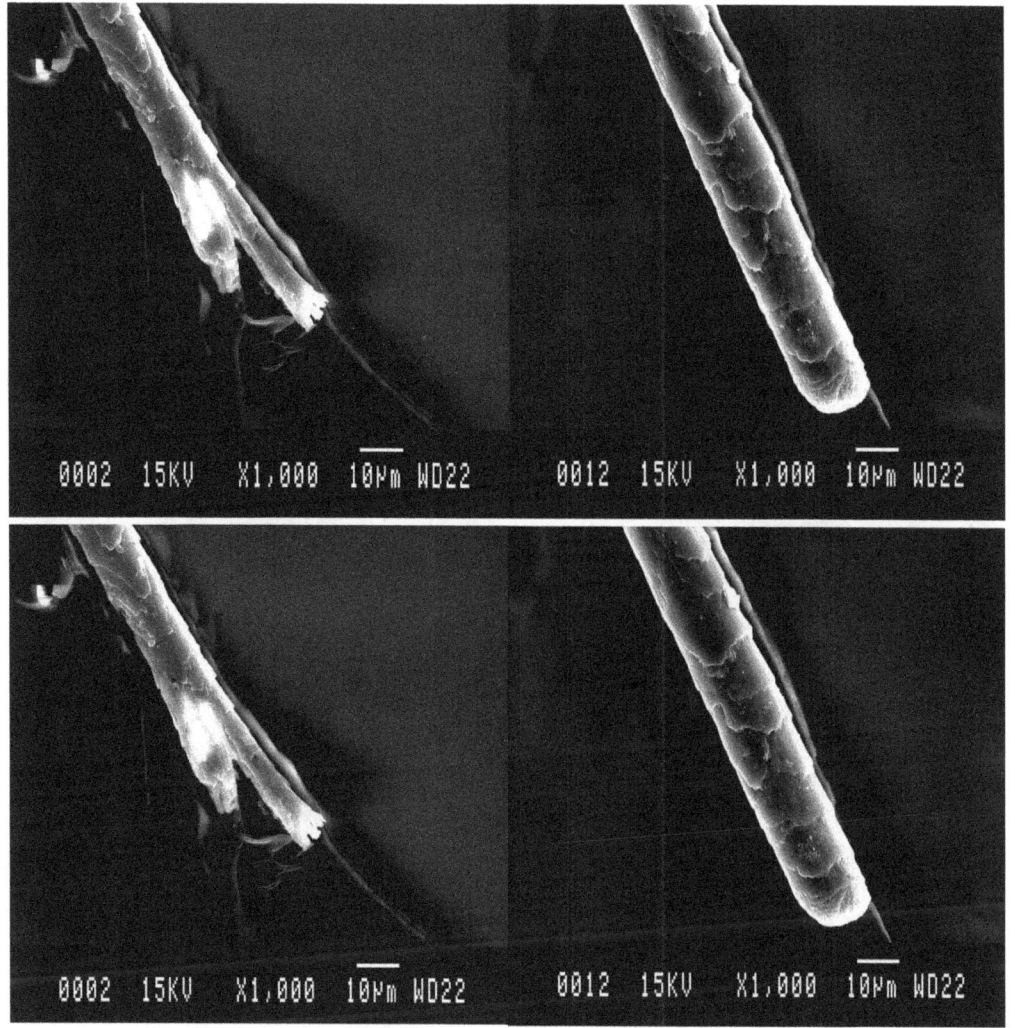

THIS PAGE AND OVER: Scanned electron micrographs [SEMs] of the hair sample found at Bolam lake. They were from a small carnivore of the cat tribe. The samples were prepared by Robert Loveridge (a technician at the University of Portsmouth), to whom we owe a debt of thanks.

He provided a CD containing some thirty images, eight of which are reproduced above., and on the next page

0003 15KV X1,000 10μm WD22
0006 15KV X1,000 10μm WD20
0005 15KV X1,000 10μm WD22

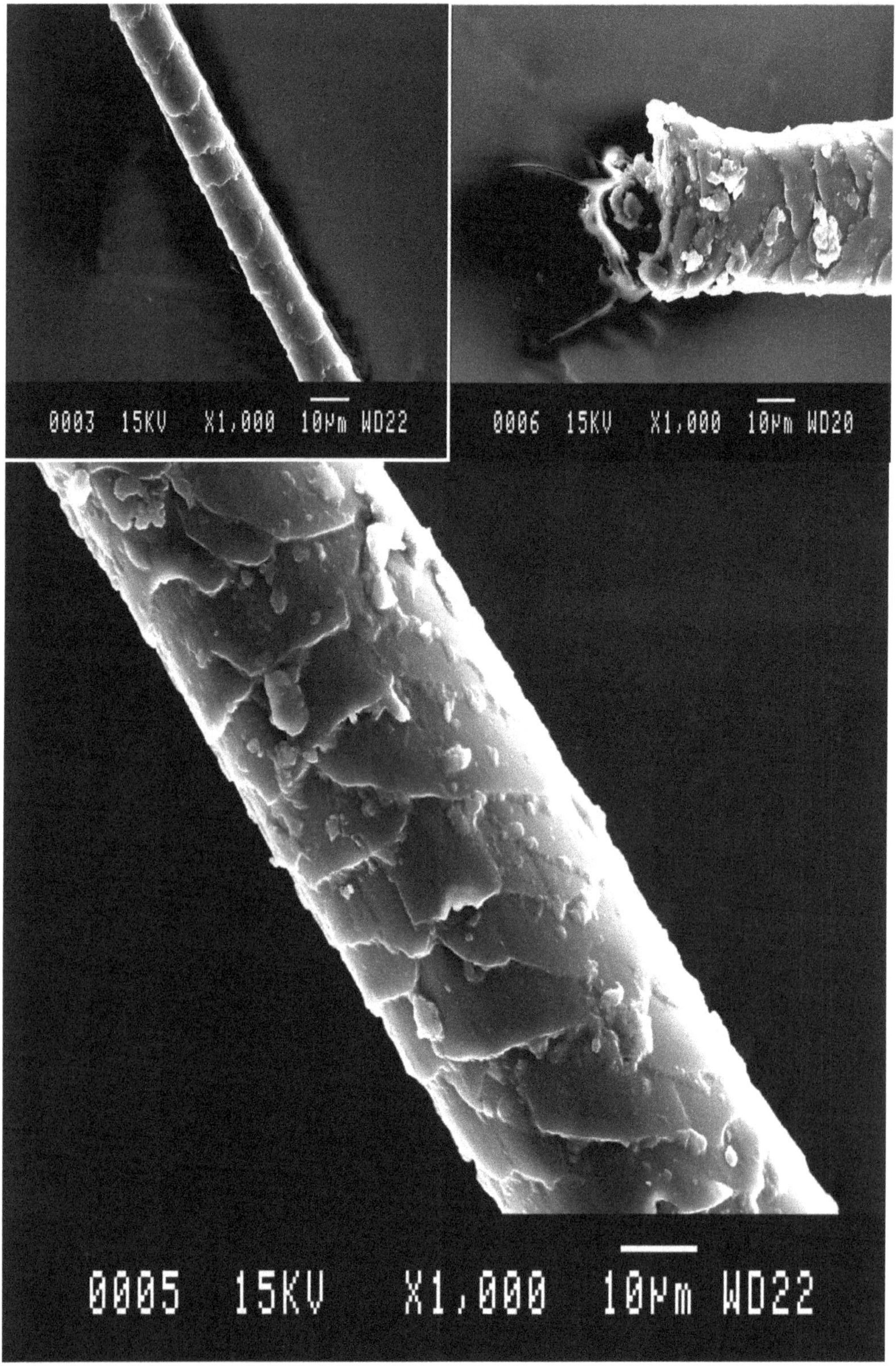

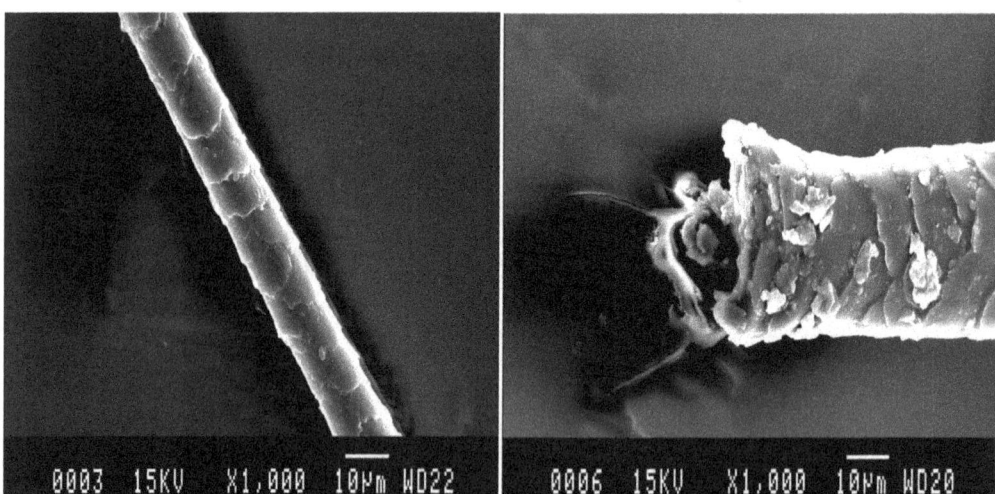

more problematical".

And he was right, of course. He also added:

"Our job is to find out the truth, whatever that is. We're totally impartial. We don't have a preconceived idea of what we're looking for. Nine times out of ten, sightings turn out to be something quite ordinary, but if we solved that mystery for future generations I'll be happy. I'm absolutely convinced the yeti exists - but not here. I think people would've found it by now. People who say they've seen the Yeti here are like people who claim Elvis is still alive and working in Sainsburys."

(I happen to know that Elvis actually runs a fruit and veg stall at South Shields Market, but its no big deal).

Teammate Richard Freeman said: *"There's no doubt that there are flesh and blood creatures still undiscovered".*

But what about the Beast? Could it be an ape? Not according to Richard:

"I actually hate apes. They are filthy, vicious, dirty, horrible things. I prefer working with reptiles. Anyway, there just isn't room at Bolam for a breeding population of apes. It couldn't happen. There simply isn't a suitable climate, habitat, or food source in Britain to support an unknown great ape, but then again I would love to be proved wrong.

We are all very excited about this trip", said CFZ Administrator John Fuller, the newest member of the CFZ Team. *"After all how often does one get the chance to confront a medieval folk tale which seems to have come to life?"*

Quite. However, although the CFZ tried to get the press to show at least a modicum of seriousness in the way they treated the story, it wasn't long before the red tops started referring to "the monkey with glowing eyes". Ah, well.

By mid-January the CFZ team were ready to travel north. Jon later recalled, *"This is not the time nor the place to give a long, imaginative, travelogue-style description of what was essentially a very tedious journey up the M1. However, we arrived in Seaham late on Thursday 16th, and after a few pints at The Dawdon Miners' Welfare Club across the road from our lodgings, we collapsed into our beds and prepared ourselves for the first proper day of our adventure.*

Serendipitously, we were able to stay at a house owned by our County Durham representative, David Curtis. He and his wife Joanne were absolutely fantastic all the way through our sojourn in the North. The only sad thing about our stay with them was that Davy had to work most of the time and was not able to join us during most of our adventures." The following morning Jon and his team liased with the aforementioned Geoff Lincoln, who had managed to get five of the possible six witnesses to talk to the team.

"I think it should be noted, here", Jon later recounted *"that the sixth witness is a soldier. With the burgeoning situation in the Middle East spiralling rapidly out of control, it would be completely unreasonable to expect a serving military man to be at the beck and call of four loonies from the CFZ."*

Later that day, the CFZ personnel met Geoff Lincoln, and Gail-Nina Anderson (a member of the CFZ Board of Consultants). At long last they could make their way in convoy to Bolam Lake itself. Jon later remarked that the lack of an eerie *atmos* at Bolam Lake was almost a let-down.

"It would be nice to say that we were overwhelmed with a spooky feeling, or that the genius locii of the location was in some way redolent of Fortean freakiness. But it wasn't. It was just what one would expect from a heavily wooded country park in the north of England in the middle of January - cold, wet and grey."

Almost immediately after arrival, Geoff Lincoln showed the CFZ team members three of the locations where the Beast of Bolam Lake had been sighted. The team carried out a thorough series of photographic mapping exercises, whilst Jon stayed back at the main car-park and did his best *"to fend off the incessant inquiries from the local press."* I was doing the same in South Tyneside, and will never forget receiving a call from New Zealand; a reporter wanted to know if it was true that the Beast had eaten a number of people. I told him of course it was true, and that rumours were already circulating that one of its victims was reported to be the Deputy Prime Minister. After giving him strict instructions not to tell anyone, and swearing him to secrecy, I could swear he had an orgasm. *"I must go! Cheers mate!"* he uttered ecstatically. God knows what his editor thought, but as far as I know it never made the Kiwi headlines.

After lunch, a TV crew from Tyne-Tees Television turned up and filmed interviews with the CFZ team and a small number of Twilight World members who had shown up.

John Fuller probably hit the right note when he told Tyne Tees, *"I don't know what to expect, but I've got an open mind. We haven't seen anything yet, but you never know what's out there."*

It was only after the TV crew had gone that Jon, Richard and the others realised that something rather strange was happening.

"Although we had tested all of our electronic equipment the night before, charged up batteries where necessary, and put new batteries in all of the equipment which needed them, practically without exception all of our equipment failed.

My laptop, for example, has a battery, which usually lasts between 20 and 35 minutes. It lasted just 3 minutes before conking out. Admittedly, I received an enormous number of telephone calls during our stay at the Lake, but not anywhere near enough to justify the fact that I had to change handsets four times in as many hours. The batteries in both Geoff's and our tape recorders also failed. It seemed certain that there was some strange electromagnetic phe-

nomenon at work there."

As dusk fell, the CFZ team drove to South Shields where they had arranged to meet with the *Twilight Worlds* members for a joint briefing. The gathering was held in the upper room at *Rosie Malone's,* a pleasant pub in the Market Square.

By the time they arrived, I'd already given my own briefing to the members. The biggest problem I had with some was getting them to take the whole thing seriously. One member had brought a rubber ape mask with him and was doing simian impressions at the bar when I arrived, and I was less than impressed. Nevertheless, the rest of the evening passed off without incident.

The following morning, Saturday, Jon awoke at 5.30 at the behest of his alarm. Not long after, a taxi arrived and took him all the way from Seaham Harbour to a lay-by 1,500 yards up the road from the Bolam Lake car park. Here he took part in an interview for the BBC Radio 4 *Today* programme. One thing of great importance happened during the half-hour or so he spent shivering by the side of road waiting to speak to "the Beeb".

"Just before dawn, the rooks which live in a huge colony in the woods started the appalling row which is presumably the corvid equivalent of the dawn chorus. Then just as suddenly, the noise stopped."

Jon heard a brief succession of booming noises – "like a heavily amplified heartbeat from a Pink Floyd record" - before the rooks started up again.

Jon said later, *"It is unclear whether these noises came from the vicinity of the lake itself, or were made by the incredibly Heath-Robinson set-up of satellite dishes, and recording equipment which was loaded in the back of, and on top of the BBC man's car."*

During the taxi journey back to Seaham the driver remarked on the peculiar behaviour of the rooks, and said it that although he was a country man himself and had spent his whole life living in this area he had never heard anything quite like it. No sooner had Jon arrived back at base, than it was time for the entire C F Z expeditionary force to drive to the outskirts of Newcastle where they met Geoff and a second witness in a cafe attached to a garden centre. They then continued on to Bolam Lake, just as the *Twilight Worlds* team was preparing to leave for the same destination.

One of the things that pleased Jon and the others immensely was that several witnesses had agreed to stage reconstructions of their encounters at Bolam Lake itself. *Twilight Worlds* members who arrived early helped out in this regard, and also with a whole battery of experiments and tests set up by CFZ personnel.

Days earlier, Geoff Lincoln had noted a series of apparently "artificial tree formations" similar to those "Bigfoot tipis" found by researchers in the United States. As noted previously, I've seen such formations in Louisiana, and they are indeed bizarre. One such formation at Bolam

Lake contained a tree trunk of approximately five inches in diameter, which had seemingly been wrenched out of position wholesale. A CFZ spokesperson later confirmed, *"It is unlikely that the cause was wind damage because surrounding foliage was unharmed."*

Intrigued by the presence of such a North American cryptozoological phenomenon in rural Northumberland, a team made up of *Twilight Worlds* members, CFZ personnel, Geoff Lincoln and Graham Inglis went off to map these formations and make a photographic record. They also took with them a *Twilight Worlds* member especially trained in using their own EMF meter, together with a dowser. After the electrical mishaps of the previous day, it was necessary to find out whether there were indeed any abnormal EMF fields in the area. Sadly, the result was a big, fat zero.

As the day wore on, a number of investigators were beginning to think that the Beast of Bolam Lake was going to make the event a no-show. It was not to be, for the Beast of Bolam was about to make his most spectacular entrance ever onto the world's cryptozoological stage.

At around 16.15 hrs, a briefing of all personnel was held in the car park. As the light was just beginning to fade, it was important that any remaining tasks to be completed that day were carried out promptly. Having completed everything that I had wanted to do, I told the group that I now intended to carry out an Indian ritual known as "calling in". This involves the carrying out of a ceremony designed to draw an animal of choice from its lair towards the location in which the ceremony is being held. In times past, Indian hunters would use this ritual to lure their prey towards them. Approximately twelve personnel – a mixture of *Twilight Worlds* and CFZ members – would assist Jon, and another group of volunteers, elected to stay at the car park. Jon was still dealing with media enquiries over his phone, and the rest were basically carrying out a "mopping up" exercise which involved taking care of minor jobs still needing attention. After the briefing was over, I went with my team to a clearing in the woods approximately half a mile away down a path which was well used by visitors.

When we reached the clearing, two members agreed to film the ritual, although I insisted that the film was "vision only", and did not include the sacred chants that I needed to utter. They agreed.

Richard Freeman had with him a food offering that would be used as part of the ritual. As far as I can recollect, the entire ceremony took about three or four minutes to complete. Then, with due ceremony, I laid down the food offering and retreated to the side of the circle made up by the curious onlookers. Almost immediately, I heard a faint but distinct howling noise coming roughly from the direction of the Visitors' Centre. I remember saying something like, "Did you hear that? It's coming". The crowd began to chatter amongst themselves, and then, after a minute or two, I heard an eerie stomping noise from the undergrowth. I pointed towards the trees and said, *"Its there".*

The beast didn't show itself to our group, which was rather disappointing. However, one of the CFZ team had a walkie-talkie so that he could stay in touch with Jon. Suddenly it crackled, and the member pressed a button and held the device to his ear.

"Its Jon. You lot better get your arses into gear and get over here fast...it's showed up."
Within seconds, the entire team was hurtling en masse along the path towards the car park. When we arrived, the scene was one of complete chaos. Jon was standing beside his car barking out orders. At his behest, team members were running hither and thither carrying out his commands.

"Get over there with a camera and start shooting..now!"..."You two...circle around to the right and cut off its exit. I want pictures, close-up and long shot. Hurry!"

It seemed to me that Jon was the only one keeping his head, although later, when I found out exactly what had transpired, I could sympathise with their panic. Had Jon not been so disciplined and professional, I seriously doubt that the event would have proved to be as positive as it did. I've always liked Jon, but he went up even further in my estimation that day.

My one enduring memory is seeing Debbie Proudlock race off into the undergrowth in chase of the beast. When she emerged, later, she was missing a training shoe. It had got stuck in a boggy patch and was left behind as she continued the hunt unabated.

As I was preparing the final draft of this book manuscript, Debbie reminded me of something else that took place around that time. On the first day of the main investigation at Bolam Lake, Debbie had lost her car keys. A fellow *Twilight Worlds* member had to run her back to Newcastle in his car so that she could retrieve her spare keys from home. They then journeyed all the way back to Bolam Lake so that she could gain entry to her own car in the car park.

I had left the park before this incident occurred, but the following day I was walking through the woods with her partner, Trevor, when he happened to say, *"Did Debbie tell you about the fiasco with her car keys last night?"*

I was just about to say, *"No, what happened?"* when there was a clinking noise. Trevor and I both looked down on the ground and, to our utter astonishment, saw Debbie's keys – the lost keys – lying directly in front of Trevor's feet. There was something rather eerie about this seemingly extraordinary coincidence which helped to convince me that there was something truly paranormal going on at Bolam Lake.

Anyway, back to the climax of the story: For what seemed like an eternity, the car park, the pathway and the woods were filled with running, shouting people. Eventually, everyone returned and all personnel were accounted for.

What had happened was this: About fifteen minutes after I had departed with my team to carry out the calling-in ritual, at 16.30hrs, one of the *Twilight Worlds* members, Anna Kaye, told Jon that she'd seen something *"large, human-shaped and amorphous"* standing in amongst the woods just in front of the car park. Another member told Jon that she thought she'd seen something too. Jon looked, but could see nothing. Whatever had been there, if anything, was now gone.

As the dusk gathered, at about 1700hrs, Jon again heard the raucous noise of the rooks that he had reported just before dawn earlier that day. Suddenly, once again, they fell silent and one of the *Twilight Worlds* members shouted that she could hear *"something large moving around amongst the undergrowth"*. Jon ordered all of the car drivers present to switch on their headlights and to put them on to full beam. He couldn't hear any noise in the undergrowth, although other people present did.

Exactly eight people were watching the woods intently. Five of them, including Jon, had their eyes trained on the thicket exactly opposite Jon's car. Without warning, an enormous man-shaped creature emerged from the undergrowth to the right, and then ran left before once again disappearing into the tree line. This was obviously the creature that I'd had "called in" minutes earlier, and which I'd heard stomping through the woods in the direction of the car park. Moments later, the creature reappeared and ran back in the direction from which it had first came. My feeling is that it had heard my team running along the path towards the car park and realised that it would run headlong into us. Hence it reversed course back along the route it had just traversed.

Of course, what the whole world and its uncle wanted to know, in every detail, was exactly what Jon and the others had seen. Perhaps Jon should have the first word:

"We only saw it for a few seconds then it vanished into the trees. This is the latest in a bizarre series of big hairy man sightings which have confounded zoologists and which have taken place in the UK in the last six months."

According to Jon, the beast was around 8 feet (2.4 meters) in height, 3 feet (0.9 meters) wide and dark in colour.

"I shouted for our drivers to put their headlights on and I saw something standing, much taller than me and moving along quickly indeed. What I saw was a man-shaped. It had a barrel chest and thick, muscular arms and legs. I had a very clear sighting but I saw no glowing eyes and wasn't able to tell whether or not it was covered in hair."

The aforementioned Anna Kaye, then a clairvoyant with *Twilight Worlds*, was with Jon when the sighting occurred. She described what she saw as a "huge, black manifestation", although she wasn't certain it was a "living creature" in the conventional sense.

"It could well be something more spiritual and I do know that there is an Iron Age fort at Bolam and I wonder if it could be connected to these sightings", she said.

One member of *Twilight Worlds* apparently did manage to get some photographs of the Beast. I was later shown, in very poor lighting conditions, what was purported to be them. They were close-up pictures of what initially looked like silhouette shots of half a head and maybe a shoulder. At first I remember being impressed, as it even appeared possible to see the outline of a hairy arm on one photograph, with individual hairs actually being visible. Later, under better conditions, I was able to examine the pictures in more detail. Photograph 1 was a picture of the forest floor with the lower right-hand aspect of the image obscured by a roughly rectangular silhouette. Photographs 2 and 3 were almost identical, the main difference being that the silhouette on the third image was shaped like a head. Photograph 4 also bore a silhouette that looked like a crab with its pincers extended and the tip of the two claws touching.

The first thing that needs to be said about the images is that none of them actually contain a "photograph" of anything other than the normal detritus one finds on the ground within a wood or forest; leaves, twigs and earth. The silhouette shapes are simply images of something obscuring the camera's flash. Whatever it was, it was not the Beast of Bolam Lake, as the photographer would hardly be taking pictures of the forest floor if they were aware the creature was quite literally breathing down their neck.

Further analysis indicates that the photographer quite simply photographed their own right hand. What I think happened was this: During the chase, Jon Downes was desperately trying to get the volunteers to take pictures. The photographer in question did as requested, and, in a state of great tension and anxiety, looked through the viewfinder. Seeing a dark shape in front, they naturally started snapping. It wasn't till *later* that the suggestion was made that the dark "silhouettes" on the images could have been the dreaded beast itself. To be fair to the photographer, they probably couldn't be sure what they'd caught on camera; it was dark and confusion reigned throughout the camp. Although they didn't realise it at the time, what they actually saw was their own right hand partially obscuring the lens. Proof of this lies in the fact that the camera is pointing towards the ground. If this *was* the Beast of Bolam Lake posing patiently to allow four photographs to be taken, then it must have been lying down on the ground at the time. Professional analysis of the "silhouettes" show conclusively that they are merely shadows, and digital enhancement clearly highlights a multitude of sticks, stones and leaves that are obscured by the shadow before the images are digitally manipulated.

I've since lost touch with the photographer, and so have been unable to reproduce the images here. Nevertheless, the sheer weight of eyewitness testimony demonstrates convincingly that *something* was there, whether or not it was actually immortalised on celluloid. Later, Jon, Richard and others would offer their feelings about the entire Bolam Lake experience.

"I've been a professional monster hunter for years", said Jon, *"but the thing I saw has prompted me to completely re-evaluate my world-view".*

Richard was unequivocal: *"This is undoubtedly the most important fortean zoological incident in the past half-century".*

The following day, Sunday, the entire team took time out to recharge their batteries and reflect. When the team returned to Bolam on the Monday, they conducted experiments to find out exactly how far away the creature - if it was a creature - was from the excited onlookers. Using Richard Freeman as a model, Jon made a fairly accurate estimate that the creature had been 134 feet away from him at the time of his sighting. He also estimated that the creature had run along a distance of between 12 and 18 feet.

The mystery animals of Northumberland and Tyneside

Jon told me later, *"You know, about five minutes after I saw…whatever it was, I walked across the car-park to where that woman had her sighting with her young son. I felt exactly the same thing; a sensation of intense unease. I was glad when I returned to the others in the car park."*

Later on that day, the team carried out a final photographic survey of the last two sighting locations. Along with Graham Inglis and Richard Freeman I tried to scan the era for magnetic anomalies using a pocket compass. I registered an unexplained anomaly at the location of the fishermen's first sighting. However, when the others tried to replicate this later in the day they had no luck.

One final coup was when Geoff Lincoln took some of the CFZ team to interview two further witnesses. The first was a young man who lived in the suburbs of Newcastle. Geoff and Richard visited him at his home and he told of his encounter with an enormous, man-shaped being next to a hollow tree in the woods some months previously. The incident had taken place whilst he had been walking his dog. He had been so frightened by his experience that he refused to ever go near Bolam Lake again.

After the investigation was over, the CFZ team returned to their base in Exeter with what the media described as their "thousands of pounds worth of equipment".

Jon was delighted: *"The expedition was a success beyond our wildest dreams. The most exciting thing was that five people had seen the beast at the same time - and I was one of those people."*

Richard Freeman commented that one of his colleagues was among the witnesses:

"What they saw was not Bigfoot, or Sasquatch as I prefer to call him; it was an enormous shadowy figure in the trees, more like a ghost than flesh-and-blood. In a park not far from a city centre, you're not going to get a nine-foot ape-like creature - England doesn't have the habitat to support it."

Richard believes that sightings such as this - and Scotland's Big Grey Man of Ben MacDhui and the Grey King in Wales - are of a paranormal nature. *"I don't mean that these are the ghosts of some creature which has died; I think it is more complex than that."*

Richard has travelled the globe hunting for monsters, and in every culture, the same types seem to crop up repeatedly. He refers to this phenomenon as the "international monster template", which is made up of dragons and other large reptiles; ape-like entities such as Sasquatch and even the trolls of Medieval Europe. He also includes in this category entities such as fairies, elves, goblins; giant mystery birds and the aforementioned Black dogs. *"I believe they are analogues of creatures which actually inhabited the plains of Africa millions of years ago. These were creatures which our ancestors would have encountered. We have fossil memories of these animals, and I think that under certain conditions the human brain can create three-dimensional images of these analogues."*

An interesting thought. However, there is, at least hypothetically, another possibility regarding the Beast of Bolam Lake – the awful thought that the entire affair may have been nothing more than a protracted hoax. I must nail my colours firmly to the proverbial mast here, and say that I reject the possibility outright. However, the suggestion *was* made in the media not long after the investigation drew to a close, and thus it needs to be examined.

Hoaxes are not uncommon in the world of cryptozoology. Perhaps the greatest is the notorious Piltdown Man affair. This also involved, technically speaking, a man-beast of sorts. Not long after the investigation into the Beast of Bolam Lake ended, rumours started to spread that the creature was nothing more than a human being in a suit. With distinct echoes of the alleged hoaxing of the Patterson/Gimlin Bigfoot footage, some ne'er-do-wells had apparently hoaxed the Beast of Bolam Lake sightings in similar manner.

What *really* happened was this. In the summer of 2002, two sixth-formers from a local High School decided to take part in a distinctly unusual arts project. They'd obviously seen the Patterson/Gimlin footage of Bigfoot, and decided to re-create it themselves. The students visited Kielder Forest in Northumberland after renting a brace of monkey outfits from a fancy dress costumier. According to one of the students, they hoped to recreate some of the famous footage from the USA where Bigfoot can be seen stomping away through the undergrowth. For three days the students tramped around the forest in their costumes. They described conditions inside the suits as "extremely hot", and their discomfort was made no better by the presence of legions of biting midges. Despite the insinuations that someone had faked the Beast of Bolam Woods sightings the following year, they adamantly denied it, and their vigorous statements to that effect were reported in the press. And I believe them. Still, the media love a good story.

Some of my fellow local researchers weren't happy about my involvement with the Bolam Lake investigation, claiming that I had no right to "speak to the press" without their permission. What on earth made them think that they had any jurisdiction over what I wrote in the papers or said on TV is beyond me, but as we say on Tyneside, "There's nowt as queer as folk". On 6 February 2003, I took them to task in my *WraithScape* column and urged them not to be so petty. I was later told that they had been negotiating with a newspaper in the south for their own story, but when I allowed the *Shields Gazette* exclusive rights to publish one of the photographs of the mysterious footprints I'd taken the other publication went cold on the story. They'd wanted an exclusive and didn't get it, and my erstwhile associates felt they'd missed out on a big pay-day. They were simply jealous.

Controversy still rages over the Beast of Bolam Lake, the mysterious creature that every now and then scares the bejabbers out of Northumberland's fishing folk. My personal take is that it is definitely not a creature of flesh and blood in the conventional sense. I think it is, to use Jon Downes' terminology, a "zooform" animal; a creature that has certain animal characteristics but is unconventional in the sense that it can appear and disappear at will. Where does it hail from? I do not know; perhaps, to use that old cliché, another dimension of sorts. What I do know is that it has some type of objective reality and was not a hoax or a hallucination.

Maybe it will return one day.

Chapter Five

The Dolly Pit Hell Hounds

What you are about to read is a true story. Only the facts have been kept the same to disturb the reader.

Shit happens, but it doesn't often happen like this. The story of the Dolly Pit Hell Hounds is nasty, nasty, stuff – cloying, unsettling ordure which burns holes in the psyche and makes you wonder what's coming next. Anyway, as long as you haven't just eaten, read on. If you have just eaten, then read on regardless. Just don't blame me if you end up with a Docker's Omelette on your living room carpet.

The nineteenth century was perhaps the acme of coal production in the United Kingdom, although few people envied the hard life of a miner. Take the Dust, for example. Everyone was scared of the Dust. One day you'd be fine. The next you'd be wheezing. The day after that you'd wheeze some more. Then you'd go to the doc's, and he'd say, 'It's the Dust'. Then you'd go home and tell the wife that you were knackered. She'd then tell the kids that the whole family was knackered, basically, because from now on they'd have to survive on pennies made by taking in washing and from Parish handouts.

Still, there wasn't much you could do. The Dust was the Universal Buggerer of Lungs, even now causing miners in America's deep South to sing odes about it, such as Kim Jones Dean's touching "No Black Lung in Heaven'. Miners in the north of England didn't sing songs about the Dust. They just told the family they were knackered and then rode out of town on a wave of alcohol-induced numbness.

The Dust only got to other people, though. Every miner hoped he'd be one of the lucky ones, and so there was never any shortage of volunteers to go underground. It was either that or starve. Starving killed everybody, but the Dust only killed many. Better take your chances with the Dust, then. Pits proliferated over the coalfields, and one such mine was known as the Dolly Pit near Sheriff Hill, Gateshead. The Dolly Pit, opened sometime before 1773, had a mixed reputation; it was an industrious mine with a decent output of coal, but it had also been the scene of several accidents and tragedies.

Near the Dolly Pit was an exposed shaft, which was apparently some forty feet in depth. Whereas today such a dangerous aperture would be capped out of concern for health and safety regulations, the 19th century did not have such rigorous statutes. Thus, in 1808, the shaft became the scene of a terrible tragedy when a depressed young woman threw herself into the hole. There she lay, a confused sculpture of blood, tissue, bone and cotton, until her pulped remains were fished out, straightened out (as best they could) and buried in the local churchyard. Or maybe not, as there were also rumours that her corpse was never actually recovered.

Why she killed herself no one really knows. Maybe her man had just come home and said, *'I've got The Dust and we're all knackered'*. Anyway, she died instantly when she hit the bottom. Poor lass. Not only dead, but lonely. Never mind. She'd soon have plenty of company.

Soon after the tragedy the Dolly Pit closed down, which meant that everyone everywhere was knackered all at once. When a pit closed down it was heavy business. *No more shoes for you to go to school in. We're on the Parish now.* With typical Geordie resolve, people would joke about covering their feet in lamp black and lacing up their toes. If you didn't have shoes, you could always pretend.

In those days, despite the widespread poverty, it was common for working families to keep a dog. This tradition was particularly strong in the mining communities. However, with the closure of the Dolly Pit, many men, now living in abject poverty with their families, found that they simply couldn't afford to feed a dog as well. Sadly, the impoverished miners hit upon a most awful solution. Quite simply, they threw their animals down the very same shaft that had been the scene of the aforementioned suicide.

This must sound incredibly cruel; indeed it is. However, it must be remembered that their owners may well have presumed that death would be instantaneous as soon as the dog hit the bottom of the shaft. Perhaps this seemed better than clubbing them to death or allowing them to die a lingering death through starvation.

Unfortunately, not all of the dogs died as quickly as their owners presumed. After the first few dogs were sent yelping to their deaths, their carcasses piled up and provided a soft, maggoty cushion for the living animals that followed. Diving headfirst into a growing pile of decom-

The Dolly Pit – home to an underground pack of ravenous canines (artistic representation).
BELOW: A sink-hole above an old pit shaft, similar to the one at the infamous Dolly Pit in Gateshead.

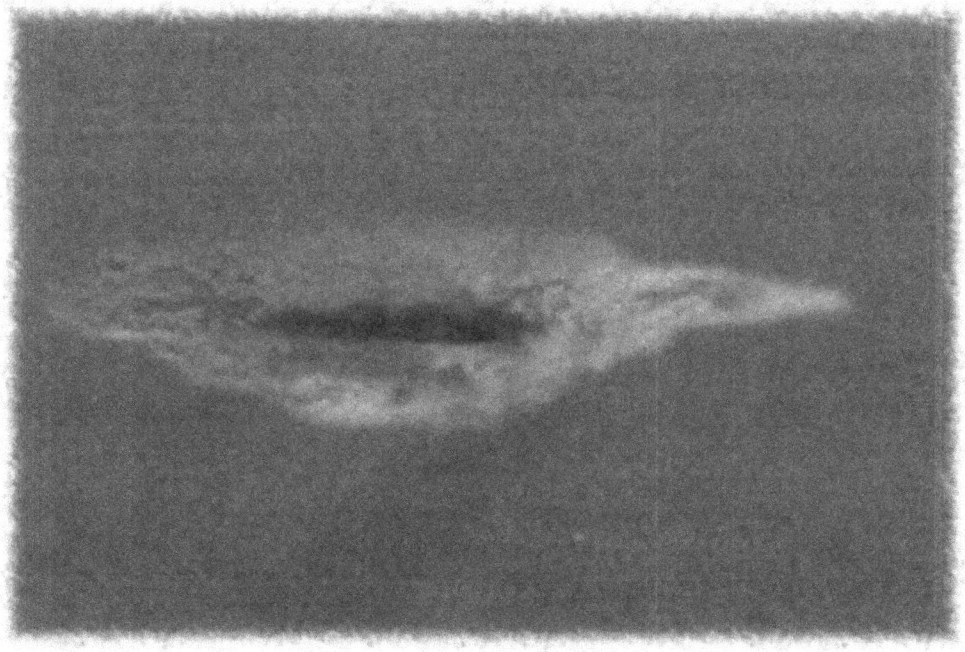

Miners from Gateshead on a day out in July 1904 – with their dog.

posing flesh is not a turn-on, even for a dog. It's a fly thing, but not a dog thing. Nevertheless, *squish* was better than *thud*, the dogs would have agreed.

The truth is that many of the animals, which survived the fall, sustained themselves by feeding on the carcasses of those who had not been so lucky. A small rivulet of water also ran through the shaft, which ensured that the dogs did not die of dehydration. Before long a colony of dogs had begun to inhabit the network of interconnecting mine-workings, and it is believed that they may even have bred. One local historian has suggested to me that upwards of two hundred animals were dispatched of in this manner, but adds that even this is "a very conservative estimate".

Against all the odds the Dolly Pit Dogs survived for several years - an incredible canine society existing, quite literally, under the feet of its human equivalent up above. However, we must assume that conditions were deplorable. The only light would have been the small amount that entered via the shaft. It must also have been incredibly cold in the winter.

The Dolly Pit Hounds would always have been hovering on the brink of starvation, I suspect. But then, just when it looked like they were knackered, someone would throw a cocker spaniel into the hole and they'd all say grace and don their napkins. Dogs don't like being cannibals any more than humans do, but hey, needs must when the Devil drives, you know.

It is said that some locals, taking pity on the animals, would throw their food waste down the shaft for the dogs to eat. Not that there was much food to waste, mind you. An old ham bone, maybe? Aye, if you could afford ham. Not many families could afford ham, especially the families that were knackered because of the Dust.

At one point a plan was devised to bring the animals up to the surface, but this was Not a Very Good Idea. It was Not a Very Good Idea because the family of the young girl who had committed suicide didn't get *her* body back, so why give a toss about some mangy dogs? This gave out the message that dogs were more important than people, and this wasn't true. Unless the Black Lung had knackered you up, of course, in which case your family would be better of with the German Shepherd. There was often talk about 'bringing them up', but it came to nought.

Eventually, the terrible practice of throwing the dogs down the shaft stopped, and one by one the animals down below were forced to turn upon each other for food. Hound would chew upon hound, dancing a macabre waltz as both animals tried to bite a chunk from the other's rump. *Rip, shriek, snap, yelp, chew,* they would have went, saliva, snot and blood mingling together as an unholy trinity of bodily fluids. One can only imagine the stench, as the shafts and tunnels became carpeted with shit and decaying carcasses.

The weakest would be killed first, of course. Or the smallest. Or the oldest. It didn't matter really, as long as there was food.

Eventually, there must have come a time when only one dog survived. This would have been the King Dog, the Meanest Dog, the Strongest Dog. Survival of the fittest, if you like. It would, at some juncture, have feasted on a lonely Last Supper of beagle, or maybe collie. No subordinates left, now. No apostles to spread the word through the Shafts of Darkness that another brother had bitten the dust and was ready to be consumed by the mourners left behind. There was nothing to do now except await a lingering death by starvation. Like those who had The Dust, the last dog down the pit was well and truly knackered.

It used to be said that, long after the last dog had perished, ghostly howls could be heard echoing around the landscape. Some say they can still be heard today, as a pack of spectral dogs roams the deserted caverns under the site of the old Dolly Pit.

I first researched the legend of the Dolly Pit Hounds in August 1999, and an abridged version of the story appeared in my weekly *WraithScape* column, in the *Shields Gazette*, that November. There is no doubt in my mind that the tale is true, at least substantially. What it says about our changing values as a society is another argument for another day. To me, it just beggars belief that anyone can take a sentient creature and throw it down a hole in the bloody ground.

The Witch Hares & Vampire Rabbits

The 'Witch Hares of Whittingham' are a very hard breed of creature to classify. It is hard to say whether they are humans who have shape-shifted into hares, or vice-versa. Or maybe something else altogether.

The concept of witch hares is not so common as it once was. Truth to tell, few people would now be able to offer an explanation of what a witch hare is, and it is likely that within the next century the idea will be found solely within the pages dusty old tomes about folklore and local history. No matter; perhaps I can help rectify the situation to a small degree.

The parish of Whittingham is to be found plumb in the centre of Northumberland, just to the south west of Alnwick. Every August there is an agricultural show held there, and the place has a deep sense of history.

Christianity has always had a tremendous influence in the Whittingham locale. Visitors may still see the Church of St. Bartholomew, the history of which can be traced back in one form or another to the mid 7th Century.

And yet, brooding just under the surface, the old ways were still to be found. Clothed as it was in a fine garb of Christian respectability, Northumbrian society still played host to various forms of paganism. Of course, there have been periods in our history when declaring such an interest could well prove counterproductive and earn you a visit from the Witchfinder General or one of his lackeys. Tolerated or forbidden, however, the earthy, indigenous spirituality of our pre-Christian era never really went away.

To good Christian folk, animals were merely dumb creatures; beasts of the field supplied in copious amounts by the Lord to provide us with sustenance. Pagans thought somewhat differently. They understood a very important truth about animals; like humans they had a spiritual-cum-psychic essence, which could be connected with under the right conditions. Because of this, it was believed by those disposed to such views that animals and humans could interact with each other in far more subtle ways than more conventional folk had hitherto considered.

Even Christians embraced many superstitions about animals, particularly in rural areas, although they would be unlikely to espouse them in front of their priest.

Around Whittingham in the 18th Century there was a belief that if a male met a woman (especially an elderly woman) on his way to work, disaster would befall him that day unless he returned home. Miners were particularly fastidious about following this rule, and miners who met a woman on their way to work would simply return to their homes for the rest of the day in the fear that if they went underground they would have a terrible accident.

At some juncture – the historical record does not denote exactly when – a similar superstition developed. It came to be believed that seeing a hare on your way to work would produce exactly the same results unless you returned to your abode immediately. In an era when staying off work automatically meant that you'd be docked a day's wages, and possibly lose your job entirely, to follow such a superstition faithfully could bring great hardship on both oneself and one's family. And yet the superstition *was* followed.

There is good evidence that those who adhered to such superstitions believed that these hares were familiars; that is, animals that lived with witches and assisted them in their magickal endeavours. Christians, of course, believed that familiars were actually demons in disguise. Witchfinders promulgated the notion that if a fly or a moth flew past a suspected witch when she was standing trial, it was all the proof necessary to condemn her. One can understand, then, how hares came to be feared by the local populace, especially after their numbers declined after changes in farming techniques. Their very rarity would make them seem all the more suspicious if they suddenly showed up.

As the years went by, these magical hares took on a negative reputation that they did not deserve. Some farmers came to see them not as conventional hares that had somehow been bewitched by the local old hag, but as a distinct species in their own right. Witch hares were not *normal* hares; they were, like black dogs, quasi-corporeal creatures that came from the depths of hell and could wreak havoc upon any humans foolhardy enough to get too close to them.

Inevitably, the delusion that had attached itself to the poor hare also infected its close kin the rabbit. White rabbits, particularly, were to be feared. Curiously, like black dogs, they were also said to have "eyes as big as saucers".

The Vampire Rabbit sits atop a door in Newcastle city centre. Who put it there and why are still unsolved mysteries.

Lucky rabbit feet – a popular totem in the north east of England to ward off evil spirits.

An unlikely hero regarding these strange circumstances proved to be a man called Edward Pease.

Pease was born in Darlington in 1767. He was the son of a wealthy wool and textile merchant who was apprenticed to his father's business at the age of fourteen. At some point Pease recognised that it would be profitable indeed if there were to be a railway to carry coal from the collieries of West Durham to the port of Stockton. Pease brought together a group of business associates for a meeting, the result of which was the formation of the Stockton & Darlington Railway Company in 1821. Inevitably Pease became a well-known and well-respected figure throughout the north east of England. What is less well known is that Pease developed something of a reputation for being a layer of ghosts and other paranormal entities.

On one occasion, a female employee went to Pease in "great tribulation" saying that she could not work. Whenever she "sat at the wheel-head" of her spinning machine a ghost would appear and frighten her away. If the woman couldn't work she wouldn't be paid, and if she received no wages then she couldn't buy food for her family. In a state of great anxiety she confessed to Pease that she was terrified they'd all starve to death.

Pease was a superb lay-psychologist. He calmly patted the woman on the shoulder and confidently informed her that he could set matters straight. With great ceremony he took a sheet of paper and, according to contemporary accounts, "dressed it with red wafers". What exactly this entailed I'm not sure, but Pease then held the paper over a fire before pinning it to the woman's spinning wheel.

Of course, the "ritual" was simply so much stuff and nonsense, but it worked. The woman visited him again in a state of elation and told him that his efforts had indeed been successful. The ghost had now vanished, and she was able to get on with her work.

I've been told that Pease carried out similar rituals in Northumberland to "lay" the witch hares that would occasionally disrupt his business. I doubt for one moment that Pease actually believed that the animals were anything other than ordinary hares, but by playing up to the superstitions of the locals, and treating those beliefs with respect, he was able to convince them that he could rid them of their malign influence.

The legend of the 'Satanic Hare' or 'Satanic Rabbit' is responsible for an unusual piece of folk history on Tyneside – and an architectural mystery.

Some years ago, an associate of my wife and I met us at Newcastle Central Station, and told us that there was an interesting curiosity he wanted to show us. We walked across the road, turned right and meandered past a number of shops. Eventually we found ourselves in a lane adjacent to the churchyard of St. Nicholas' Cathedral. The street is officially called Amen Corner, although curiously that name doesn't appear on any map in my possession. Amen Corner is a discrete little niche of offices that you would probably never see unless you had a reason to go there. You certainly wouldn't stumble across it by accident as it's off the beaten path, so to speak.

On entering the lane, immediately conspicuous is a terraced building currently occupied by a firm of solicitors. The edifice - a delightful example of Victorian architecture - is fronted by an ornate entrance that is a delight to behold. I'm normally against painting stonework; the results are almost always hideous and can transform otherwise beautiful architecture into garish monstrosities that have all the taste and subtlety of a clown's hat. I'll make an exception in this case, however. The frontage has been decorated in a style reminiscent of Romanesque, and the colours of pink and white, although on paper quite hideous, actually hang together pleasantly and are very easy on the eye.

Above the door is a delightful circular window surmounted by a semicircular pelmet. And it is here where the pleasantry stops and the mind is catapulted into an altogether darker frame, for above the pelmet crouches the brooding 'Vampire Rabbit of Newcastle'.

At this juncture I must tell the reader that I can offer no explanation for the presence of the rabbit. No one knows why it is there or what its significance may be. There has to my knowledge only been one suggestion made, and I'll deal with that presently.

The vampire rabbit sits staring - exuding an aura of menace. With it's bulging eyes and bared fangs one could be forgiven for thinking that it was contemplating pouncing upon the unwary observer.

Despite the fact that the origins of the vampire rabbit are an enigma, there is an old story that the rabbit is actually a hare. Several years ago, during renovations to the building, the creature's ears were accidentally broken off and replaced. Unfortunately, they were replaced the wrong way round, making the beast's appearance even odder than it was before.

From time-to-time, Internet chat rooms and message boards receive postings from locals offering suggestions as to what the function of the vampire rabbit was intended to be. Two years ago, a poster suggested that the creature was some sort of protective totem designed to a) shield the residents of the building from harm, and b) act as a warning to those who would enter therein with evil intent.

There may indeed be some truth in this, although I can't help but feel that the creature is connected with the Northumberland witch hares in some way. The old legends about the witch hares are slowly fading into the fog of history, but at least the vampire rabbit still stands, keeping guard, a testimony to days gone by.

Chapter Seven

The Gast Adders of Northumberland

Adders (*Vipera berus*) are one of two snake-like animals found in the wilds of Northumberland. But then again, strictly speaking, they are the *only* true snakes found in that locale, as the other, the slow worm (*Anguis fragilis*) is actually a legless lizard. The grass snake - or ringed snake - (*Natrix natrix*) and the smooth snake (*Coronella austriaca*) are not found in Northumberland – only further south.

Adders are not plentiful in the north of England, but although their numbers are declining they are not likely to be in danger of extinction. Their habitats are reasonably well protected and are to be found in the remoter areas that have not yet succumbed to the relentless march of so-called civilisation. One good place to spot an adder would be in the area around Kielder Forest.

Adders are not aggressive by nature. In fact, they are quite placid provided they are not disturbed, threatened or alarmed. They like to bask in the sun, and to the observer may appear quite docile. This would be a misguided assumption, however, as the adder has a sophisticated defence mechanism that rivals that of many other species, and it will not hesitate to use it if needs be. Over the epochs of development that it has enjoyed at the behest of Mother Nature, the adder has perfected a system for delivering venom into the bodies of its victims that is chilling in its efficiency. However, adder bites are rarely fatal, and for that we may be thankful.

In April 2007, during a family outing to Shell Bay in Studland, Dorset, 11-year-old

Michael Chapman was bitten by an adder as he ran through the heath. His leg swelled up and turned vivid purple, and Caroline Davis, who had taken Michael on the trip with her own family, thought the youngster was going to die. After being rushed by ambulance to Poole Hospital he was given morphine and anti-venom treatment and made a full recovery. A doctor at the hospital later admitted that his condition had been, essentially, life-threatening.

April – indeed any time during the spring – is statistically the worst time for adder bites because the rise in temperature brings them out of hibernation.

Also known as the Northern viper, the creature derives its name from its birthing habits. Viper is a combination of two Latin words, *vivus* (live) and *parere* (giving birth). Together the words, as a collective noun, mean something like, "live-birthers", and speak of the snake's habit of giving birth directly to living young instead of laying eggs like most snakes. The name "adder" is derived from the Old English *Naedder*. The Welsh names for the adder are *Neidr* and *Gwiber*, although there is some evidence that in ancient times the Gwiber was an entirely different creature altogether; a flying serpent similar to other cryptozoological winged snakes revered by Native Americans.

It would seem logical to assume that the reverence given to the adder in Northumberland folklore is due to the fact that it is the only snake that *could* be revered, as it has no competitors. This notion doesn't paint the full picture, however, as the adder was believed to possess some distinctly unusual properties.

Despite their ability to deliver sometimes-fatal doses of venom into their victims, adders generally have a positive image. They are perceived to be filled with wisdom and knowledge, even if a little sly at times.

The difficulty we are faced with when studying the adder in Northumberland is that, as in Wales, there seems to have been two distinct species that bore the name; one, the taxonomically classifiable species accepted and well known to science, the other a more arcane creature with distinctly zooformic attributes and tendencies.

Adders like to live in the open countryside and prefer rough land covered with heath and coarse shrubbery. In the spring, adders shed their old skins and prepare to breed. Warm temperatures excite adders into a sexual frenzy, and males will fight ardently for the rights to a female of their liking. In Northumberland it was believed that adders used to dance. We now know that this isn't true. What was interpreted as dancing was simply a larger, aggressive male attempting to intimidate a smaller, weaker rival in the mating game by engaging in some impressive gymnastics.

Female adders are fussy creatures, and will travel considerable distances to find a suitable place to birth. After several months of gestation, adders normally drop their young in late summer.

Like salmon, adders have a liking for home and always return to the original hibernation site.

If left undisturbed, the site may serve generations of adders over the years. Adders are adaptable and can cope with most impediments that come their way. However, they dislike the cold intensely and fare badly in harsh winters.

Adders have a varied diet and will pretty much eat anything if they are hungry. Given a choice, though, they prefer to feast on mice, frogs, rats, lizards and small birds. Adders are efficient hunters and can strike with lightning speed. Rather callously, they will then rest and watch their victim as it slowly succumbs to the venom coursing around its system. When the prey is dead, the adder will swallow it whole before the process of digestion begins. It should be noted here that the fluids within the adder's digestive tract are extremely potent and will even digest bones quite easily. The only undigested body parts you're likely to find in adder poop are teeth and hair.

We've taken a brief look at the conventional, taxonomically recognised adder and its lifestyle. However, there is no distinct line to be drawn between this and its more "paranormal" counterpart, which I once heard referred to as the gast adder. (Gast is the old Anglo-Saxon word for ghost, but whether this is the correct derivation of the word in the context of adder nomenclature I wouldn't like to say.) When studying contemporary literature dealing with adders, it is often hard to make a distinction between the two, as superstitions and legends were often applied to both.

For example, there is an old Northumbrian legend, also found in other parts of mainland Britain, that female adders will often swallow their young to protect them from a perceived threat and then disgorge them later when the coast is clear. Biologically, it is difficult to see how adders could do this, as the digestive juices within the female are so strong that they would kill the young very quickly indeed. It has been suggested that the adder's unusual procedure of giving birth to live young may be responsible for this superstition. The belief is that, at some past time, an adder must have been killed and cut open. The hunter may have found some live young inside the mother's body ready to be born and assumed, because they were not encased in eggs, that the mother must have swallowed them after birth to protect them from some unknown prey. There is a problem with this theory, however, even though it is ostensibly very persuasive.

Because the adder is the only snake to live in Northumberland, it is hard to see how locals could have been able to compare it with any other species that gave birth to its young by the laying of eggs. Of course, it isn't impossible that ancient Northumbrians could have been aware of other species that did lay eggs, but it is hard to believe that they did *not* know that their own, native snake gave birth to live young. I think, therefore, that we must be cautious here and not jump to convenient conclusions. An old farmer from the Scottish Borders once told me that he had seen a mother adder swallow her young, although here again we must be cautious, as adders have been known to eat each other if they're feeling peckish. Was the adder he saw feasting on smaller snakes, the offspring of another female, or was it really swallowing its young as per the old tales from yesteryear?

Alongside the conventional flesh-and-blood adder there seems to have existed a more spectral creature of the same name. This creature was also believed to swallow its own young to protect them, but it was also believed to cast spells on those who stumbled across it before luring them to a remote area where it could devour them at will. Of course, snakes of all types have long been accused of "hypnotising" their prey, and although some naturalists deny that this happens there are numerous accounts of other creatures – particularly rabbits – being mesmerised before the snake strikes and kills it.

The more zooformic adder is believed to live for hundreds of years. Other stories relate that it cannot be killed. These are probably connected to the old legend that adders "never die until the sun sets", and this in itself needs some explaining.

Adders are hardy creatures and can survive, even if only sometimes temporarily, the gravest of wounds and injuries. In Northumbria there used to be a belief that if you did manage to strike an adder a mortal blow it would not die immediately, but would slither under a stone and wait till the sun had set before finally expiring.

It was also believed that adders could provide powerful protection to those who knew how to solicit it from them. There were ordinary adders, and there were those of a more occult nature that possessed the legendary *Glain Neidr*, or "serpent glass". These adders were not of the flesh-and-blood variety, but the herpetological version of the black dog; creatures that may look like conventional snakes, but were really something different altogether.

There is a belief stretching back to Britain's pre-Christian, Druidic era that adders would, every spring, gather in a huge council to select the Adder King for the coming year. This congress would inevitably end in a huge battle, in which rivals for the adder throne would fight with each other. After the battle, and the new king had been selected, the congress would be dispelled and the snakes would go on their way. However, they were believed to leave behind a large patch of foamy froth as evidence that the congress had taken place. The Druids believed that if you looked carefully in the middle of this foamy substance you would find a precious stone called the *Glain Neidr* or 'Adder Stone', sometimes also referred to as the *Maen Magl*. In Northumberland the stone was known colloquially as an "adder's egg" or an "adder's stone", and at least one village is named after the notion.

Descriptions of the stones are remarkably consistent. They were said to be perfectly spherical and almost transparent. In colour they could be azure, terracotta or pale green. Some were said to be pink or lilac in hue.

I have actually seen a number of adder stones, and the first matched the traditional description perfectly. At first I was of the opinion that it looked like a highly polished blue marble, but there were two unusual features to it that made me wonder whether I was indeed looking at something distinctly odd. Firstly, I noticed that if I looked through the stone from one side it was totally opaque. However, if I turned the stone around and looked through it from the opposite side it was perfectly transparent. There may be a scientific reason for this, but I cannot think of one. The other thing I noticed was that if I held the stone in my right hand it felt distinctly warm to the touch. However, if I moved the stone to my left hand it felt icy cold, as if it

A sign pointing to the Northumbrian village of Adderstone. BELOW: A hedgerow where Adder Stones were once found.

A field near Adderstone where an Adder Congress was alleged to have been held some years ago. BELOW: A glass Adder's Egg, owned by Morgan Martin.

Another 'Adder's Egg' photographed by the author in Hexham.

An adder or viper *(Vipera berus)* the only poisonous snake known to exist in the United Kingdom. It is widespread throughout western Europe, and is found all the way to eastern Asia. Its bite is seldom fatal.

had just been removed from a freezer. The owner of the stone said, "*Ah, you must be left-handed then!*" I confirmed that this was indeed the case, and he commented, "*Adder Stones are often like that; cold in the hand you use to write with, and hot in the one you don't.*"

I asked the owner where he'd got the stone from, and he told me that a friend of his had found it at an "Adder Congress" near Rothbury many years ago, and had given him the stone as a memento shortly before he died. After the adders had departed, he went to inspect the scene and, sure enough, the ground was covered with a thin, milky-white foam. Remembering the old tales about adder stones, he fished about and found exactly what he was looking for.

Adder stones were once prized for their healing and curative properties, particularly in cases of chronic infection. It was also believed that they could render the owner invisible, and that adders themselves could simply disappear as long as the stone was in their possession. Curiously, they were believed to keep the stone balanced on the top of their head. There are old tales about adders "disappearing in a flash", but the question that needs to be asked is whether it was possession of the "mystical stone" that gave the adder the power to render itself invisible, or whether there is another explanation.

In the weird world of paranormal research I learnt along time ago "never to say never". What you ridicule and take the piss out of today may well turn out to be true tomorrow. I will refrain, therefore, from casting scorn on the notion that adder stones can render their owners invisible. However, a far more likely explanation for the creatures' ability to appear and disappear at will is that it is not the flesh-and-blood adder that can do this, but its spectral counterpart, the zooform adder.

Adder stones had a darker side, however. Although they could be used for healing, they could also be used to inflict harm on enemies. Surreptitiously dipping an adder stone in the ale of someone you didn't like would, it was believed, transfer the poison or "sting" of the adder to the unwitting recipient. Conversely, wearing the stone as an amulet would protect the owner from harm just as efficaciously.

Adder skins were given pride of place amongst diviners in Northumberland, and were used by some when important decisions had to be arrived at. The skin would be cast to the floor, and its position would be carefully noted and used to predict the future.

There were two trees that grew prolifically in Northumberland that were intrinsically linked to the adder stone legend. Hazel trees were believed to attract adders that were still in possession of their stone. Find an adder next to a hazel tree, then, and you were guaranteed to find an adder stone perched on its head. Ash trees had just the opposite effect. The magic of the ash was said to be too powerful for adders, and the creatures would never go near one.

King Arthur was, according to legend, killed in the Battle of Camlan in 537 AD. The battle was said to have been precipitated by an adder that appeared between Arthur's army and that of Mordred. One of Arthur's men pulled out his sword to kill the creature, but Mordred, believing that this was a sign that Arthur's troops were beginning to attack, took the initiative and charged first. No one knows where the Battle of Camlan took place, Wales being perhaps the most likely location. However, there is also a theory that it may have taken place in Northumberland.

The reader will have gathered by now that there is a common theme to the mystery animals of Northumberland and Tyneside. For each conventional, flesh-and-blood creature there seems to be a spectral or psychic counterpart, which looks the same superficially, but is radically different in its *modus operandi*. There are dogs, and there are black dogs. There are rabbits, and there are giant rabbits and were-rabbits. There are swans, and there are ghost swans. There are hares, and there are witch hares. To this list we may also add adders, for there are ordinary adders and gast adders.

Like black dogs and witch hares, it was believed that the appearance of gast adders were a bad omen. If an adder crossed your path, then the amount of bad luck that would come your way would more than compensate for any good luck brought to you by a black cat. The spontaneous appearance of a gast adder near your house – or worse, inside on the hearth – was a sure sign that someone close to you was about to shuffle off this mortal coil and go to meet their Maker.

There is a curious parallel between one aspect of Northumbrian adder lore and that of Native Americans across the Atlantic.

In the tradition of some Native American peoples, dreams can be influenced by the creation of a "dream catcher". Dream catchers can be used to protect oneself from bad dreams, precipitate the arrival of good dreams, etc. The dream catcher is essentially a hoop containing an intricate "spider's web" fashioned from thread or cord. In the centre of the web is a hole that will allow good dreams through, whilst a small dream stone attached to the web near the hole filters out bad ones.

The style and design of dream catchers differs from tribe to tribe. Some replace the web with a simple cross-fashioned from two pieces of cord. Others do not use a hoop at all, but simply a stone with a hole in the centre.

Stones with holes in the centre often have great value to Indians of differing tribal backgrounds. My aunt, who was born and raised in England and knew little about her Indian heritage, still kept a stone with a hole in it on her hearth. Hole-stones had to be formed naturally, and not fashioned with tools.

In Northumberland and County Durham there was a second tradition regarding "Adder Stones", but which was markedly different to the one discussed previously.

It was long believed in these counties that the "sting" of an adder was so powerful that it could burn holes even in rock. Hole-stones – or holy stones, as they were oft called – were believed to have been formed when an irate adder struck out and stung the stone. Such stones were believed to contain powerful magic. These stones were essentially identical to those used by the

American Indians, and often employed for identical purposes. A holy stone could be hung above your bed or your window (exactly the same as in Indian tradition) to "catch" bad dreams. They were said to be especially efficacious in preventing what is generally known as "Old Hag Syndrome"; a deeply unpleasant symptom of several sleep disorders, including narcolepsy.

It is strange indeed that in some parts of Northumberland holy stones supposedly created by adders were used to ward off bad dreams, and yet there seems to have been a parallel tradition in the same localities that dreaming of adders meant that your enemies were plotting against you.

Few people in Northumberland talk about gast adders now, but many still hang on to the old superstitions regarding conventional adders. Some say they bring good luck, others believe they portend death and destruction. I choose to believe the former, as is my democratic right, and can testify to the power of both Native American hole-stones, and Northumbrian adder stones. I use both. Its what you call hedging your bets…

Chapter Eight

The Cleadon Panther

I've investigated many mystery animals in my time, but having one present itself virtually on my doorstep was indeed a pleasant surprise.

I live within the village of West Boldon in the County of Tyne & Wear. Next to West Boldon is East Boldon, and then just a bit further towards the coast again is Cleadon Village.

For those readers who are unfamiliar with the finer nuances of the Geordie geographical outlay, Cleadon is a village that lies 'twixt the City of Sunderland and the town of South Shields. And it was here, on the morning of 11 January 1999, that the notorious 'Cleadon Panther' first reared its bewhiskered head.

If you believe the stereotypes – and I don't - Cleadon is "different". Cleadon, you see, is where the money people live. The locals ride horses, continually talk about Gavin and Fiona doing something in the city and have a strange compulsion to drop the phrases 'on the Algarve' and 'Must remember to pick my Ferrari up from the garage' into every other sentence.

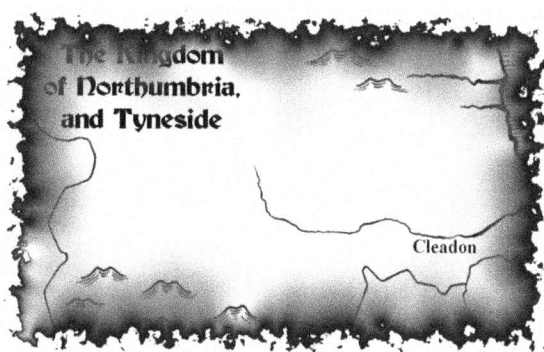

Before Cleadoners start sharpening their very expensive kitchen knives ('When I'm cooking I refuse to use anything but Ganichiu, dear; you can only get them from good stores and they'll set you back a mere £300 a set') I must remind them that I do *not* share these negative conventions. As I live near Cleadon I can vouch forth that they are actually a canny lot. True, many Cleadoners are extremely well-heeled, but a friendlier bunch of locals you'll not find anywhere. Cleadon folk may often be rich, but I rarely find them pretentious. Cleadon's problem to some, though, is that although it is an extremely nice place to live it is not at all sexy. To those without taste, Cleadon will be akin to beauty without wilderness, casserole without salt. They're wrong, but that's the impression and there's not much we can do about it.

Broadly speaking, the history of Cleadon Village is well documented. Pick up a copy of Hodgson's *The Borough of South Shields*[1], and the facts about Cleadon are there for the imbibing. But there is more to Cleadon than facts. Walk away from the neatly manicured lawns and genteel shops and you will sense a strangeness. The hills are alive in Cleadon, oh yes. Civil War ghosts still cut the night in these parts, and no one can explain why birds rarely sing near the abandoned WW II gun emplacements.

To add to the high strangeness of Cleadon, we would do well to remember that the 'Black Helicopter Phenomenon' has also manifested itself here. I know of two witnesses who have seen fleets of these unmarked aircraft silently cruising through the sky towards God-knows-where.

But that's another story; to me, Cleadon is one of the brightest jewels in South Tyneside's crown. Visit it and you won't be disappointed.

Anyway, I digress. Back to the big cat. The first I knew of the Cleadon big cat was that evening when I read Iain's article.

Iain Smith, a journalist with the *Shields Gazette*, had been tipped off about the creature's arrival and was promptly dispatched by the editor to interview witnesses. Locals were truly disturbed when one woman was frightened by the feline interloper whilst making a sandwich, and something obviously had to be done.

The second witness, a man who did not wish to be named, told Iain that he saw what looked like a very large cat running past the back of his dwelling at about 12 noon. According to the man it was completely black, had a small head and a long, curled tail. The witness was sure that what he saw was a big cat of some description as it ran past a gap in the hedge.

As you do, of course. Another witness informed Iain that he was something of an amateur naturalist and had been a keen twitcher for 35 years. He also admitted to being a sceptic who, had he been told about the "panther" by others, would simply not have believed them. Seeing something with your own eyes is a different matter, however, and our birdwatcher has now joined the ranks of the true believers.

The witness told Iain that he wasn't really sure what the creature was, but because of reports in

---

1. Hodgson, George B: *The Borough of South Shields* (South Tyneside Libraries, 1996) ISBN: 0 906617 25

the press he had immediately thought of a puma or a panther.

Iain's story appeared in the *Gazette* the following evening with the headline, **CLEADON BIG CAT IS ON THE PROWL!**

In the article he recited the experiences of several other residents who had also seen something large, dark in colour and cat-like loping past their hedges. Large, dark and cat-like, eh? There are three large, black cats living on our estate alone. One of them, which belongs to my son Martyn and his wife Rachel, is called Misty. Misty likes nothing better than to piss on our rosemary bush, but hardly deserves a write-up in the papers.

"*No*", Iain told me. "*We're talking really big, Mike. This is no domestic cat*".

One chap who lived in Woodlands Road said that what he saw was definitely a puma or panther. Another resident from Moor Lane saw it and immediately phoned the police. Enough, already. Something was afoot, and it was by all accounts a rather large one with retractable claws.

At once a wonderful ambience of paranoia took hold, and Charles Fort would have just loved it. Retired colonels started polishing their blunderbusses, and a local resident allegedly demanded that the SAS be drafted in. Cleadon, a relatively sedate village which had hitherto resisted adopting the nick-name Excitement Central, suddenly had a crisis on its hands.

At 0530hrs the next morning, I put on my best spandex tights, donned the rest of my Crypto-Man costume and flew off to Cleadon Village. Walloping Wildcats, Robin, have the good people there not suffered enough?

Cleadon Village at dawn on a January morning is not a fun place to be. It was snowing heavily as I walked past the duck pond towards Cleadon Lane. Ah, the whiteness of it. The landscape had an altogether Dickensian appearance, but where were the rag-clad orphans? Whither the match girls, chestnut sellers and purveyors of lavender posies? God, those bastards just don't make Victorian vignettes like they used to.

The steely silence was only broken by the odd hoot, croak, rustle or other noise unidentifiable by us townies. If the panther was lurking, I'd find him there.

At 06.10 precisely I happened to be standing at the fence of a field on the east side of the lane. In the field were four ponies. Three stood together, whilst the fourth ambled around some distance away. Suddenly the lone pony whinnied and galloped over to the others. All four were now agitated, and continually shuffled around. To the north I caught a glimpse of a large, dark shape beside the hedge. It seemed to be creeping towards the animals a few feet at a time. One of the ponies whinnied again, and the silhouette disappeared into the hedge. Was it the big cat? Or perhaps a black dog? I do not know, but it certainly scared the horses, as they say.

There is a common belief that wildcats like nothing better than to roam around the countryside looking for humans to devour. This is nonsense. Unless a wildcat is threatened, demented or very, very hungry it will keep well away from humans. This is why, despite hundreds of sightings of mystery big cats over the years, only a handful has ever been caught.

Some researchers have suggested that up to eight viable breeding populations of wildcats are now ensconced within mainland UK. Conventional wisdom has it that these breeding groups are made up of big cats that were released into the wild by their owners, after they were unable to comply with new government restrictions concerning the keeping of exotic pets introduced in 1976. Rather than have them put down, they let them go. One of those groups, it is rumoured, is situated in Co. Durham at a place called Kiln Pit Hill. As pumas are solitary animals, could one have made its way north from Co. Durham – a short distance – and taken up residence in Cleadon? I scoured the White, looking for both inspiration and an answer. No flower girls, no pickpockets, no muffin men. Just the rapidly deepening White and four ponies in a field opposite the village school.

I spoke to the caretaker of a local school. Had she seen any economy-sized felines prowling around the vicinity? No, but she'd certainly heard the rumours. Later in the day, Iain went back to Cleadon Hills with a *Gazette* photographer for a photo shoot. Just for fun he dressed up in a pith helmet and khakis and carried a rather formidable looking hunting rifle which would undoubtedly have got him detained by the local constabulary had they caught him with it. In fact, it looked like it had been filched from the set of *Zulu!*

Looking every bit like Colonel Fawcett, Smith posed for some magnificent snaps. The Cleadon big cat could now be either a Fun Thing or a Serious Thing, depending on your mindset.

Within an hour Iain Smith was already working on another story for the *Gazette*, and he did his editor proud. With all the gravitas of the sober journalist he penned, with the imaginative byline of Iain "Safari" Smith, a carefully crafted appraisal of the facts under the headline, "Here, Kitty, Kitty!" The "large" cat had now gotten larger, one resident describing it as being twice the size of a large dog.

Smith's article was liberally festooned with photographs. There he stood on Cleadon Hills wearing the aforementioned pith helmet and kakhis and carrying that enormous rifle. Things were hotting up, it seemed.

The following evening the *Gazette* received a call from someone who believed she could solve the mystery. The lady concerned, from Seaburn, told *Gazette* reporter Zoe Burn that she regularly walked her dog around Cleadon hills. Her canine companion was a Burmese mountain dog and had an imposing physique. Could he have been mistaken for a panther? Unlikely. He is multicoloured, long-haired and totally the wrong shape. Plus, I think it highly unlikely that his owner would take him walkies to Cleadon Hills at 0530hrs in the middle of January.

By Friday, 15 January – just five days after the story broke – the merriment and jesting were dying down somewhat. Iain Smith penned another article with the more sober headline, "Why

I Won't Walk My Dog on the Hills". It told the story of an elderly Cleadon couple who had given up walking their Labrador on the hills after the husband encountered the 'Cleadon Beast' – as it was now called – both by the old Water Tower and the abandoned Mill House. According to the man's wife her husband was not easily scared. However, his encounter had left him badly shaken.

After Iain published their story a tension settled over Cleadon. Cleadon may be posh, but think not that the residents have no belly. Make ye no hasty judgement that they have gone soft, yellow or lily-livered. Just the opposite; Cleadon's residents know how to pull together in a crisis, and, as more reports flooded in, *this* crisis looked as if it was going to be gold-plated economy-sized one. Children were frightened to go to school, women were wary about walking to the shops. Suddenly it just wasn't bloody funny any more.

Yet another resident suggested arming the Neighbourhood Watch with air rifles, whilst another demanded that the army be called in. Some local youths talked about setting up a "patrol" and arming themselves with baseball bats. When an elderly woman, sobbing, claimed that the beast had chased her across the road, what had started off as a bit of a laugh was now starting to look like some serious business.

By the following day, January 16, the Cleadon panther was no longer a Fun Thing. Iain Smith interviewed one chap who had stumbled upon the beast one morning whilst walking his dog. According to the man's wife he was really shaken up when he returned from his ramble in the hills. He vowed never to go back there again as he'd seen something which really frightened him.

This sighting precipitated a mass boycott of Cleadon Hills by dog walkers, ramblers and twitchers. People genuinely became frightened for their pets, their children and themselves – although not necessarily in that order. By January 17, the whole subject had taken a far more serious tone. The police were getting concerned, and rumours started to circulate about a rabbit being found mutilated. Try as I might, I could not substantiate the story. At 10 o'clock, Iain Smith telephoned me and asked if we could compare notes. Within two hours we were drinking tea and talking business.

The *Gazette* devoted a whole page to the beast on January 19, in which Iain put forward my theories and those of other pundits who felt they had something to offer. Iain Smith's article – *Is There Really Anything Out There?* - took a sober look at the several explanations that had been tentatively suggested.

Iain consulted me over the contents. I had suggested to him that he contact a sergeant from the Durham Constabulary. The officer was then a Durham Police wildlife officer, and had an above average knowledge of big cats. As Durham had been (and continues to be) the scene of numerous anomalous wildcat sightings, he was amply qualified to speak on the subject. According to the sergeant, there had been similar sightings in Burnmoor, Consett, Plawsworth, Durham, Castle Eden, Weardale, Wingate and Barnard Castle. I had already suggested to Iain that a rogue wildcat could have moved to Cleadon from any of these areas, and showed him several routes that the creature could have taken. Iain put my theory to the police officer, who said it certainly wasn't impossible.

The article was accompanied by a map of the area detailing numerous sightings of the beast. Personally, I felt that a rogue puma from Durham was the most likely explanation. This may have provided some clarity to the villagers in terms of what exactly they were dealing with, but it didn't settle their stomachs. What it did do was persuade more witnesses who had previously sat on their stories to come forward.

Incredibly, it turned out that villagers had been catching glimpses of the beast during the previous eighteen months, but hadn't said anything in case they were thought to be crazy. Percy Maddison told me, "*Look, its okay when a few people have seen things, but you don't come out with this sort of thing on your own, do you? Can you imagine what they'd say in the club? They'd say I was a bit mad*".

Quite. But Percy wasn't alone. Albert Burr outed himself as a witness, as did others.

These reports stoked up the Fear Factor incredibly, and Cleadon Hills were eventually declared off-limits to walkers until the Beast was caught. The headline *Big Cat Strikes Fear Into Hills,* loomed large in the *Gazette* of January 21st.

By then the fear factor had really taken hold. Iain reported that another local beauty spot had almost become "off-limits" to walkers, and the idea that the big cat was simply a large dog was firmly put to bed. The aforementioned Albert Burr, a resident of Hawthorn Avenue, spotted the beast several times whilst walking his two Rottweilers. Hard on his heels came a resident of Whiteleas, who came across the beast's footprints whilst out walking with a friend. According to the woman there had been three or four of them in the mud, and they were huge. She also stated that the prints of the paw pads could clearly be seen in the indentations.

Meanwhile, headlines such as **CLEADON BIG CAT NOT FOUND YET** and **CLEADON MAN DOESN'T SEE PUMA** still excited the intestines of the local populace. But there, I'm jesting again, and I shouldn't be. The story was about to take another serious turn, but more of that presently.

Another Cleadon resident cast doubt upon the zoological expertise of one of Her Majesty's constables after spotting the creature meandering around the village.

"It was the size of a Great Dane, it had huge, rippling muscles and its mouth was filled with the most wicked-looking teeth I've ever seen. The policeman asked me if it couldn't have been next-door's tabby cat. He must be f*****g bonkers."

Absolutely, but by Wednesday 27 January, the altogether more taxonomically aware police sergeant from Durham declared that he had information which led him to believe that several pumas had been released into the wild by a private collector in the 1970s. Apparently someone

Cleadon Village – invaded by a mystery big cat in 1999 – and again in 2007.
BELOW: The ruin of the old mill house on Cleadon Hills where the Cleadon Panther was said to "hang out" during the daytime.

Cleadon Mill. Several dog walkers claim to have encountered a large, black cat near the abandoned mill house. BELOW: Two mysterious shafts of light photographed inside an abandoned gun emplacement on Cleadon Hills at the height of the search for the mystery big cat. Other witnesses reported similar anomalies – some of them visible to the naked eye.

had released four wildcats into the Weardale area and the idea that these could have been responsible for some of the sightings was indeed a logical one.

Officially, the British Isles do not play host to wild cats, at least not of the outsized variety. The Scottish feral cat is really just a domesticated cat that's been working out in the gym, and doesn't exactly strike fear into the heart of our friends north of the border.

Readers in the US, mainland Europe, Africa, etc. should remember that the British Isles are infinitely smaller than their own continent. We have no wildernesses to speak of. Just about everywhere is within sight of a cottage, roadway or - more likely - concrete jungle filled with urbanites. Because our remote areas are so much smaller, it is relatively easy for two wild cats to find each other and start to breed. There is now little doubt that this has happened, and that - like it or not - the puma has thus become part of our *fauna*.

On February 20, things took a novel twist. I found another witness who had seen the creature at close range, but incorporated some curious details into her account.

At 8am On Sunday, 10 January Mrs. X had left her house in Laburnum Grove to purchase a newspaper. As she reached the bottom of her street and was about to turn right into Cleadon Lane, she suddenly decided against buying the paper, remembering that a friend of hers had one delivered by her son. She was a bit concerned about her friend anyway, as she hadn't heard from her for a few days, and thought it best to give her a call. She would, she reasoned, have a cup of tea with her friend and read the newspaper whilst she was there. (I still haven't worked out whether Mrs. X's primary motive was to be a good neighbour or save herself 75p, but I'm happy to let the good Lord above settle that one.)

To get to her friend's house, Mrs. X had to walk along Cleadon Lane - this section of which is actually called Front Street - and turn right into Sunderland Road. She would then turn right again into Whitburn Road, where her friend lived. She didn't get that far. After turning into Sunderland Road, our witness was amazed to see a huge, black "panther-like" creature standing on the pavement and looking across the road. Without hesitation it then "licked its lips and padded across the road to the other side. Then it suddenly disappeared."

The weather conditions in this case will be judged crucial by some. It was, says Mrs. X, "slightly foggy", and inevitably cynics will pounce on this as an explanation for what she told me next. The "panther", said Mrs. X, looked perfectly normal except for its legs. Even though she could see the rest of the creature with crystal clarity, its lower limbs were totally invisible. I asked Mrs. X if she had an explanation for this, and she replied, "*No. It must have been the fog. That's all it could be, isn't it?*"

Mrs. X also said that the creature had "bright red eyes", and that these frightened her. When asked if the eyes were *glowing*, she said, "*I'm not sure that 'glowing' is the right word, but they were very bright*".

And what about the sudden disappearance of the creature? Did it just vanish into the fog, perhaps?

"No. It was just like turning the TV off. It just sort of went out in an instant."

I carried this witness's story in my weekly *WraithScape* column for the *Gazette*. Again, I suggested that the creature might have been what cryptozoologist Jonathan Downes has called a "zooform" animal; that is, one which has the general appearance of a normal creature but which is not "real" in the common sense of the word. Zooform animals sometimes appear with limbs missing, including, bizarrely, the head and/or legs.

Zooforms are never caught, although they can, enigmatically, leave physical traces of their presence such as foot or paw prints. The "red, glowing eyes" are another common feature, as previously discussed, and have been seen in incidents involving black dogs and were-rabbits.

And then there was the puma poo. Strange turds smelling of sulphur were reported to by sitting in nearby fields, patiently awaiting examination. Alas, heavy rain carried them gracefully into the soil and beyond the reach of scientists and intrepid reporters alike.

Other unusual witness reports began to convince me that the animal might indeed have been zooform in nature. A witness said that on one occasion the puma changed colour, going from tan to black in a few seconds.

I told Iain Smith that the animal definitely would not be caught and that it would, in a few weeks, disappear. I also told Iain that he could quote me on that if he liked and if I were wrong then I'd put my hands up and take the flak.

By March 19 the hunt for the "Cleadon Panther" had become a Fun Thing again. The newspapers were talking about "Big Cat Mania", and the South Shields Museum and Art Gallery staged an exhibition called "Claws" dedicated to both prehistoric and modern big cats. Hordes of people from South Shields, Cleadon and the Boldons queued up to observe the display, desperate to see close-up what it was that they most certainly *didn't* want to see close-up whilst walking Fido on Cleadon Hills.

The exhibition opened with a Fun Day, and the entire world and its uncle wanted to get in on the act. Whilst would-be big game hunters were organising themselves into expeditionary forces to go searching in "them thar hills", the RSPCA was giving out advice to the public, the Cat Protection League was recruiting volunteers and a Northumbria Police Wildlife Officer (they didn't want to be outdone by their rivals in the Durham Constabulary) was giving still others the benefit of his wisdom.

Of course, a Fun Day wouldn't be fun unless there was something there to traumatise the psyches of little children, and in this case it was a life-size, moving replica of a sabre-tooth tiger attempting to rip the innards out of a poor little baby mammoth. The kids loved it, of course, and could now understand why both Daddy and Fido would have been very upset if they'd bumped into THIS mother-of-all felids out on the hills. Total hysteria was now the order of the day.

In April there was a momentary wobble of the enthusiasm indicator, and Iain Smith even speculated in the *Gazette* that the cat might have gone further south.

But then the public's collective juices started surging again when reports of other mystery big cats started coming in from Devon and Cornwall. Helicopters were whizzing around using the latest tracking gadgetry in an effort to find the beasts, and one or two Cleadon residents asked why we couldn't have the same degree of assistance.

A week later, there appeared to be a break in the story. Sergeant Bell had again heard that a private collector had released a number of pumas into the wild, and suspected that the Cleadon beast could be one of them. Iain wrote up the story – *Fresh Lead in Mystery Big Cat* – in which Bell appealed for the collector to come forward. All the police wanted were the facts. How many big cats had he released, where and when? To my knowledge the man never came forward.

By May, Big Cat Fever was showing no signs of slowing down. The by now legendary Durham Constabulary Wildlife Officer came to the museum to give a talk on the whole subject of mystery big cats. People were so enthusiastic he had to stop back and give another one.

I attended the first lecture, and although I found myself disagreeing with one or two minor points in the main it was solid stuff built on sound research. During the discussion it was mentioned that a certain town in Canada was "good puma country" and yet no one ever saw them. In fact, there are more sightings of Bigfoot recorded than those of pumas. This statement would be eerily prophetic, considering that two years later the 'Beast of Bolam Lake' would firmly take hold of the cryptozoological baton in the north east of England.

After his lecture the officer joined me in a question and answer session. We both tried to placate anxious residents who were still concerned that they might be eaten by the beast.

When I got home I checked my e-mail. There was a threat from someone calling himself `The Phenix` who promised to *"rise from the ashes and destroy"* me. My sin, apparently, had been to label the Beast as a carnivore in one of my own articles.

"Everyone knows that pumars [sic] are vegitariens [sic] and won't hurt a fly. Its sick idiots like you who make peoples [sic] want to turn against them and not stop doing things. Watch your back. I will strike you when you think I wont [sic] and you will not know nothing about it till it is too late to do nothing. Be warned".

And then, without warning, the bubble seemed to burst. The big cat had gone. But then, just weeks later, according to some reports it was back again. More accurately, perhaps it could be said that it never went away. Those who followed the exploits of the beast during its first incarnation were by no means certain that it ever left the area. One local businessman whom I spoke to told me, *"I've seen the thing several times. A few of us round here have, but we wouldn't tell anyone. They'd just shoot the thing."*

Well, they could try. Shooting a spectral cat isn't easy, though. I've been told that trying to fill a zooform beast with lead is an awfully risky business. According to conventional wisdom, you just hear a pop, a flash, and then get deluged with a shed load of negative psychic energy - fallout which may just be enough to send you off into the Great Beyond yourself. Don't try this at home, kiddies. Another two other witnesses told me the Beast had *"glowing, red eyes...really bright, like torches"*.

Naturally, the *Twilight Worlds* Paranormal Research Society, to which I then belonged and which had its centre of gravity in nearby South Shields, was intrigued. Some members believed that it might have been possible to make the beast manifest itself using several arcane spiritual rituals. At the time I was sceptical about this, but my experience later at Bolam Lake made me realise that the scientific approach is limited and that such rituals can indeed play a part in cryptozoological investigations.

We intended to discuss plans at the forthcoming Management Committee meeting for an expedition to Cleadon Hills in the hope that we could get the beast to appear. Then? Then we planned to photograph its hairy ass a thousand times before it disappeared into the ether again, that's what.

I think at the time I was pretty pissed off with zooforms nancying around and then buggering off before they could be scrutinised. It's been like this for centuries. Well, no more, we thought. We were mad as hell and we weren't going to take it any more. Using fair means or foul, we were going to corner the Cleadon puma and snap a few for the family album.

Well, at least that was the plan. For some reason it never came off.

Two days later The Penis, as I had fondly came to call him, e-Mailed me for the second and last time. He'd changed his mind about doing me in. Instead he was going to let Pantharon do it. Pantharon is, apparently, the Deity of Vegetarian Panthers – at least within the twisted confines of Penis's cranium. Pantharon was pissed at me for reasons already outlined by Penis in his previous electronic epistle. Sadly, he never did show up, which just goes to show that vegetarian deities are usually pantyhose - wimpo – mineral water drinkers who should keep their mouths shut. Give me carnivorous deities every time. I mean, can you imagine Jehovah asking his followers to slaughter a low-fat nut roast for Passover? I've never heard from Penis since. Maybe he's gone back to his alternate universe which is inhabited by phenixes and vegetarian pumas.

What are we to make of it all? That a wildcat of some description was roamin' through the gloamin' cannot be disputed. There were just too many credible witnesses. There is no doubt in my mind that Cleadon village did play host to an anomalous feline of some description, and it may indeed have been a puma, or perhaps a melanistic leopard, which had travelled from one of the creature's known hunting grounds to the south.
The problem is that proving the reality of the beast is not necessarily to understand its essence.

The mystery animals of Northumberland and Tyneside

Was the Cleadon puma a *bona fide* animal, or something far stranger? My instinct draws me to the latter conclusion.

So then, where does Mr. Downes fit into the picture? Well, it was The Big Fellah himself who taught me about *zooform* animals – animals that are not "real" in the accepted sense of the word, or at least they aren't real all of the time. Zooform animals seem to be able to appear and disappear at will. Whilst here they can interact perfectly well with the physical world as we know it, but they can, seemingly at a whim, slip into some alternate reality.

Having originally been convinced that the Cleadon big cat was actually nothing more than a puma that had been released into the wild, I slowly came to the conclusion that this explanation certainly didn't fit all the sightings recorded in that area. No, I think one of Mr. Downes' zooforms paid us a visit back then and I said so publicly in my column. Despite the fact that huge numbers of people were marshalled to search for the beast, I knew that its paranormal nature would make capture quite impossible.

On 26 February I told *Gazette* readers, "*Only time will tell which category the Cleadon Beast belongs to, but I will put my neck on the line now and make a prediction: Whatever the creature is, it won't be caught*". And of course, it never was. Eventually the sightings died down and my Crypto Man costume was put back in the trunk where it remains to this day. The Neighbourhood Watch has been stood down, Iain Smith has taken off his khakis and, once again, dogs and people can be seen ambling around Cleadon Hills. Well, at least most of them can. One or two still avoid the area in the belief that something dark and terrible clings there, waiting. I think they're wrong. I e-mailed the venerable Mr. Downes and the honourable Mr. Freeman and invited them to come up and investigate, but, alas, they had already made plans to visit more distant climes. However, from the day I invited them the Cleadon beast has never been seen since.

Personally I think Pantharon wimped out and couldn't face up to the Big Fellah. Pantyhose veggie deities…they're all the bloody same.

Seriously, though, if Cleadon is a truly Fortean place, then perhaps we should look for a Fortean answer to the mystery. I still lean towards the idea that the beast was – is – a zooform creature. It is neither of this dimension nor fully of the next. The soul of the Cleadon puma hangs between heaven and earth, defying logical explanation.

Almost as fascinating is the incredible effect that the beast's presence had on the minds of the general public. The cat acted like social glue, rallying neighbours and stirring passions. It was the wartime spirit all over again. The episode was also a fascinating study in Jungian psychology, for, love the mystery big cat or hate it, it presents us with a primal archetype; *man versus beast, and the triumph of man..*

On reflection I suppose we may not have seen the last of the Cleadon big cat. As soon as the next large paw print is found they'll be hanging out the bunting and dusting down the painted wagons. And good luck to 'em, I say. It was great while it lasted.

Chapter Nine

The Laidley Worm of Spindleston Hough

Three miles to the west of Bamburgh Castle in Northumberland lies Spindleston, home of "the wyrm of Spindleston Hough (sometimes called Heugh, – "heugh" being an ancient colloquialism for a ridge or raised neck of land).

The story of the 'Spindleston Wyrm' has been told in many forms, and a vicar of Norham – seventeen miles away – once penned a poem in the creature's memory.

The story centres around a royal dynasty centred at Bamburgh Castle in Northumberland. Historians may differ over the minutiae, but we know enough of the history of those times to be able to put some literary flesh on the bones, so to speak.

The presiding Anglo-Saxon monarch at Bamburgh, in the setting of the tale, is normally referred to as Ida, Idda or Eda. Indeed, there *was* an Ida connected to Bamburgh - a warlord of Anglo-Saxon stock who had disembarked with his soldiers at Flamborough Head, Yorkshire, in or about the year 547 AD.

There was a touch of destiny attached to this historical vignette. Ida's own grandfather, Ossa the Knife (or Ossa the Knifeman), had landed on the same spot decades earlier and conquered in battle no less a person than the legendary British King Arthur.

Not wanting to have his momentum checked, Ida and his troops began a forced march northwards towards Bamburgh. There was good reason for this. Ida knew that if he was to consolidate his position he could not do so alone; he would have to unite the disparate tribes, bands and clans into a consolidated kingdom that would minimise the risk of him being overthrown by someone else within a short space of time. Ida achieved his ambition, and witnessed the birth of the Kingdom of Bernicia; the forerunner of the powerful Kingdom of Northumbria.

Ida, we know, passed away in the year 559 AD. As Ida was alive during the historical setting of our narrative, we can state confidently that the remarkable incident in question must have occurred within that twelve-year period.

Ida had somewhere in the region of twelve sons - not unremarkable in those days. Six of them were born to his wife and at least another six to a number of concubines. Due to the rather uproarious nature of the times, we cannot be certain about the identity of the wife who plays such a pivotal role in our story. We need not be too concerned over this. However, what we do know from available records is that Ida's main wife passed away and he remarried. Probably around the year 550 AD, he set out to find himself a new spouse who would help him govern his burgeoning household.

The king had a number of daughters, too, one of whom is referred to simply as "Margaret". In all probability she would have been called by one of the more European versions of the name popular with the Anglo-Saxons, such as Margarete, Margaretha or Marguerite. During her father's absence, Margaret was left in charge of the castle and longed for her father's return. To her satisfaction, Ida eventually returned to Bamburgh with his new bride.

Ida had a rather lax view of marital fidelity, and from the beginning of the relationship his new wife felt distinctly uncomfortable about this. As the months passed by, her disquiet became something of an obsession, which turned the king's offspring against her and made the queen herself hate them in turn with a passion.

In particular she detested her step-daughter, for no other reason than that the princess was the offspring of one of Ida's previous wives. This alone seems to have soured the relationship between the two women, but it was an incident that occurred when the new queen had first arrived at Bamburgh, which destroyed any hope of cordiality between them.

When Ida first returned to Bamburgh with his new bride, the chieftains of all the Anglo-Saxon bands along the Borders greeted the couple; a normal enough procedure for the return of the sovereign and the new woman in his life. Naturally enough, Margaret pushed her way to the front to greet her father; after all, she was the king's daughter. Innocently, Margaret attempted to cement good relationships with her new stepmother by telling her that everything at Bamburgh was now hers and that she would personally endeavour to make the queen's life as easy as possible. One would imagine that Ida's new bride would have been delighted at this welcome, but the unfortunate interjection of a local chieftain wrecked things completely. The warlord, whose name is unknown to us, had never seen the young princess before and was quite taken aback by her charm and beauty. Without thinking, he exclaimed, *"My lady, I believe that you surpass all women in beauty and in stature"*.

The road to Bamburgh Castle – and the legend of the Laidley Worm of Spindleston Heugh.

BELOW: The eerie Bamburgh Castle – this ancient edifice lies at the heart of one of the north-east's most enduring wyrm legends. (Picture courtesy of Darren W. Ritson).

A spectacular picture of Bamburgh Castle at sunset (courtesy Darren W. Ritson).

BELOW: The steep walls of Bamburgh Castle formed an almost insurmountable barrier for invaders.

The coast to which the Childe of Wynde returned to rescue his sister from the Laidley Wyrm.
BELOW: The village of Spindleston Hough – the cave where the Childe of Wynde defeated the Wyrm can still be seen there to this day.

One of the alleged entrances to the cave system which played host to the Laidley Wyrm.
BELOW: Close-up

Another alleged site of the "cave" at the top of Spindleston Heugh.
BELOW: The path allegedly taken by the Childe of Wynde as he made his way to the lair of the Wyrm – now flanked by a dry-stone wall.

The Laidley Worm.

This kind statement was undoubtedly uttered with good intentions and seems to have pleased Ida and his daughter no end. It did not, however, please the new queen.

"You might at least have made an exception for me!" she snarled at the chieftain with venom.

From that point onwards the queen had Margaret set firmly in her sights as a rival that would have to be eliminated at the first opportunity.

It seems that the new bride of Ida was adept at using Anglo-Saxon witchcraft, and – if the tale is to be believed – worked her evil upon young Margaret and transformed her into a "laidley [loathsome] wyrm". Margaret, as one can imagine, was none too happy about this. Distraught, she apparently slithered into a cave at nearby Spindleston Hough and took up residence there on a permanent basis.

Whilst all these shenanigans were going on at Bamburgh, one of Ida's sons, known in literature as "the Childe of Wynde", was taking care of some business in Europe. According to some account he was "engaged in foreign wars", but it makes no difference.

It's perfectly understandable that Margaret should become miffed at being transformed into a monster, and it seems that, during her wyrmhood, she took to venting her wrath in the local community; you know how it is; razing the odd homestead to the ground, eating a few cows, etc. The Childe of Wynde, concerned to hear that Bamburgh was being terrorised by such a creature, immediately made preparations to travel home. What the Childe of Wynde did not realise was that it was Margaret – in wyrm form – who was causing all the bother.

Whether Margaret's brother suspected that there was more to the story than met the eye, or perhaps had an inkling that witchery was involved, we do not know. However, instead of simply hiring a boat to take him home, the Childe and his men build their own ship with three masts carved out of rowan trees, to which they added sails made from the purest silk. According to legend, the Childe and his compatriots then returned home with haste.

On their arrival back in England, the crew recognised the distinctive silhouette of Bamburgh Castle, and quickly made for the shore. Unfortunately, the evil queen saw the ship from a tower and recognised its peculiar handiwork as that of the Childe of Wynde. The queen assumed, quite wrongly, that the Childe must have discovered what she had done to his sister and was terrified that Ida would discover the truth. In an effort to stop things spiralling out of her control, she then sent out three [or in some versions two] evil companions known locally as the "witch wives" to magically attack the craft and send it to the ocean floor. Their plan was foiled by the fact that the masts of the ship were fashioned from rowan wood, which according to ancient lore protects sailors against all forms of malignant chicanery.

The queen was furious, of course, but seems to have had a "Plan B" up her sleeve. Hastily she dispatched an armed gang on another vessel with the sole intent of sending the Childe of Wynde to the afterlife. The prince, however, was no mean swordsman and, with the help of his friends, promptly dispatched his aggressors into the world beyond instead.

Whilst the battle raged just off shore, Margaret's personality was slowly but steadily being consumed by that of the wyrm. In a state of dementia, she was unable to ascertain that the sailors in the ship were actually her brother and his trusty comrades. According to the legend, the Wyrm then physically attacked the ship and kept driving it back into the sea with its tail.

The prince was naturally angry that, as well as witches and vagabonds, he now had the Wyrm itself to deal with at so early a juncture. He made a tactical withdrawal and surreptitiously put his vessel ashore further north at Budle Bay. All the while, still not realising that the wyrm was in fact his sister, he determined to track it down and to kill it with his sword.

At some juncture the Childe of Wynde found the wyrm and steeled himself for the final conflict. He withdrew his blade and made ready to dispatch the creature forthwith. However, just as the prince was about do battle, the wyrm spoke: *"Put your weapons away, Child Wynde, and kiss me once, twice, three times. I am full of venom it is true, but I do not wish to harm you"*.

After a little persuasion – heroes are not normally in the habit of kissing wyrms - the Child of Wynde pecked the creature three times as requested and the wyrm promptly retired to the farthest reaches of the cave. This act seems to have precipitated a change in the beast, for, seconds later, the transformed Margaret came forth restored much to the amazement and delight of her sibling. The couple then made for Bamburgh Castle, where King Ida was more than a little relieved to find that rumours of his daughter's demise had been inaccurate.

But the Childe of Wynde was not yet finished. As the wickedness of Margaret's stepmother became apparent, he determined to find her and dish out her just desserts. According to the tale he promptly transformed her into an ugly toad and declared that she would henceforth hop around the environs of Bamburgh Castle from then till doomsday.

Of course, this story may just be what it seems; a romantic fairy tale devoid of any historical substance. However, from time to time sightings of a phantom toad, which spits angrily at young women before disappearing, still trickle in.

Paradoxically, much of this strange tale is entirely factual. Bamburgh Castle exists. There was indeed a King Ida and, yes, he did have sons, daughters and several wives and concubines. Spindleston Hough? The place exists, and the infamous cave (or several locations claiming to be the cave) can still be viewed to this day. In fact, Jackie and I visited the village in September 2007.

I had assumed that the cave would be easy to find, but I was mistaken. One villager told me that the Wyrm had lived not in a cave at all, but a well. He gave directions as to how to reach it, and off we went. On the way, other locals fed us a number of different stories. Indeed, the only thing we could determine with any accuracy was that there are a number of locations in the vicinity where the Wyrm allegedly took up residence.

Just next to the caravan site at Spindleston there is a large depression, which appears to have been filled in with concrete and tarmacadam. I was told that this was the site of the well where the Laidley Wyrm lived, although a walker also told me that it was nothing more than the site of a banana slide that once stood there to entertain the kiddies.

A more likely site is a nearby set of crags on a gently sloping hillside. One version of the Laidley Wyrm story claims that the cave entrance one sat in one of these crags, and was subsequently demolished or collapsed. This may indeed be the case, and a fence, which borders a farmer's field, was said to roughly parallel the route that the Childe of Wynde took when he marched to the cave to confront the Wyrm. How anyone could know this with any degree of certainty is beyond me, however.

Piles of stones can be seen in some of the fields around Spindleston Hough, and someone also mooted that these may mark the spot where the cave (or well) entrance once stood. I can't believe it. These small boulders look as if they were dumped in their current location weeks if not days ago, and show no signs of embedding themselves into the landscape.

As with all the bizarre creatures within the pages of this book, the primary question that must be addressed is whether the Wyrm too really existed in some shape or form. Fantastic though the story is, it would be impetuous to discount the more paranormal aspects of the tale completely. This part of the story must, I would venture, have *some* factual basis. The shape-shifting capabilities of the Wyrm place it firmly in the hitherto explained category of zooform animals and it may just be that the Childe of Wynde interacted with one of the most bizarre cryptozoological creatures in English history.

Chapter Ten

The Shony

Centuries ago, during the Danelaw [1], the north east of England was governed by the Vikings. The Norsemen were brave fighters, infused with the darkling spirit of the warrior ethic and driven by the hope of glorious Valhalla to come. They were also a bunch of nutters, incidentally, and had some rather unpleasant habits, which would have very likely precluded their membership of a modern day Rotary Club or Tea Dance. Looting, murdering and deplorable table manners are three that immediately spring to mind.

But although it is true to say that the Vikings were fearless in battle, they did have an Achilles heel. They were, for the most part, incredibly superstitious and spent a lot of time appeasing various and sundry deities. They were also rather wary of the Shony, and so spent quite some time appeasing it as well. In between the raping and pillaging, that is.

The Shony was a monster which lived in the North Sea, and used to hang around ships when it was stormy, waiting for the odd sailor or two to get washed overboard. Its victims were not killed straight away. The Shony would take them down to an undersea cave made of coral and

1. The Danelaw, as recorded in the Anglo-Saxon Chronicle (also known as the Danelagh; Old English: *Dena lagu*; Danish: *Danelagen*), is an historical name given to a part of Great Britain in which the laws of the Danes held predominance over those of the Anglo-Saxons. The part of Great Britain which was part of the Danelaw is now northern and eastern England. The origins of the Danelaw lie in the Viking expansion of the 9th century, although the term was not used until the 11th century. With the increase in population and productivity in Scandinavia, Viking warriors sought treasure and glory in nearby Britain.

From Wikipedia, the free encyclopedia

imprison them there until it was time for lunch. Some sailors would do the decent thing and try to rescue their friends. This pissed the Shony off quite fulsomely, and to demonstrate his ire the monster would simply imprison the second sailor in the coral prison and toss the first one back into the sea to drown. No wonder they did quite a bit of appeasing, those Vikings.

But the Shony was not easily appeased, no sir. To keep this psychopathic leviathan sweet, the Vikings played a game, which involved shoving your hand into a bag and pulling out a stone. Whoever pulled out the wrong colour would then have to appease the Shony. Unfortunately the rules were quite strict about this. You couldn't just toss a few coins into the water and keep your fingers crossed like they used to in the good old Roman days. Money was of little use to the Shony, for there was no Burger King franchise at the bottom of the North Sea to spend it at. Instead, the captain would simply slit your throat and bunk you over the side. And that was that. There was nothing like a good old bit of appeasing to liven up the day.

Personally, I do not think that the Shony was a myth. A legend, yes; but not a myth. The Shony is, I would venture, a real, live *bona-fide* sea creature which still roams its old hunting grounds today. Fair enough, I don't buy the bit about the underground gaol made of coral and all that jazz, but there is, I think, enough evidence to put up a *prima facie* case for its existence. Let us take a look at it.

Holy Island lies off the north east coast of England, and is home to both Lindisfarne Priory and Lindisfarne Castle. In the 12th Century, an unusually large number of corpses were washed up on the island's beaches, much to the consternation of its inhabitants. The bodies appeared to have been horribly mutilated, one unusual feature being that the eyes were nearly always missing.

Consternation turned to terror when several people reported seeing a huge, nut-brown coloured creature swimming off the coast. The Shony, terror of the Romans and the Vikings, had returned.

By the end of the 13th Century, the number of corpses being washed up began to wane. Eventually, the locals began to wonder if the monster had really existed at all. Perhaps the strange wounds on the bodies had simply been caused by contact with the jagged rocks that could be found off shore. Maybe natural predators such as crabs had removed the eyes. (Even in those days there were rationalists and sceptics to contend with.) Whatever the truth, the terror of the Shony slowly but steadily evaporated, until there was nothing left but a vague sense of unease.

As detailed earlier, I write a weekly column in my local newspaper *The Shields Gazette*. The *Gazette* is the oldest provincial newspaper in the UK, and has never been afraid to speak its mind throughout its history. (Incidentally, in the 1920s a *Gazette* reporter was one of the minor players in the Colonel Fawcett mystery, and may actually have stumbled upon the answer to the explorer's strange disappearance, but that's another story).

One of my column's regular readers is Malcolm Urquhart, who has on occasion fed me with some good material from his own extensive archives. Years ago he handed me a yellowing,

The eerie and enigmatic Marsden Bay – a familiar haunt of the Shony.

One of the many caves that can be found along the cliff face at Marsden. Some are no more than small niches, whilst others connect to larger caverns and tunnels deeper in the rock face and even below ground.

One of Marsden Bay's larger caverns.
BELOW: A narrow fissure leading into the cliff face at Marsden.

Marsden Rock. According to legend there is a subterranean chamber beneath the rock (the south side of which has now collapsed) which may be connected to the Shony legends.

BELOW: A later photograph of Marsden Rock. Notice how the top of the rock on the south (right-hand facing) side had worn away considerably in between the two photographs being taken.

Marsden Rock after the collapse of the southern pillar.

BELOW: Some of the worn stone blocks that made up the Roman quay at Marsden.

More remains of the quay. BELOW: Marsden Rock has assumed such iconic importance in the area that it has become a popular subject for paintings by local artists.

A photograph from the cliff top of what may be the remains of the Roman quay at Marsden. BELOW: The same image with the possible outline of part of the quay highlighted.

Part of the rebuilt West Gate at the Roman Fort of Arbeia, South Shields. BELOW: The front of the Marsden Grotto pub as it appeared in the year 2000. A complex network of tunnels and caverns lies both behind and beneath the inn.

Peter Allan's grave at nearby Whitburn churchyard. BELOW: The Lambton Wyrm as depicted on a pillar inside the Cave Bar of the Marsden Grotto. Above the head of the Wyrm is the image of Sir John Lambton who allegedly slew the creature.

A close-up of the head of Sir John Lambton and the head of the Wyrm. Below/next page: Images of successive Lords of Lambton who suffered when Sir John refused to heed the counsel of a wise woman when battling with the Wyrm.

A stylised dragon head also carved on a
pillar inside the Marsden Grotto.

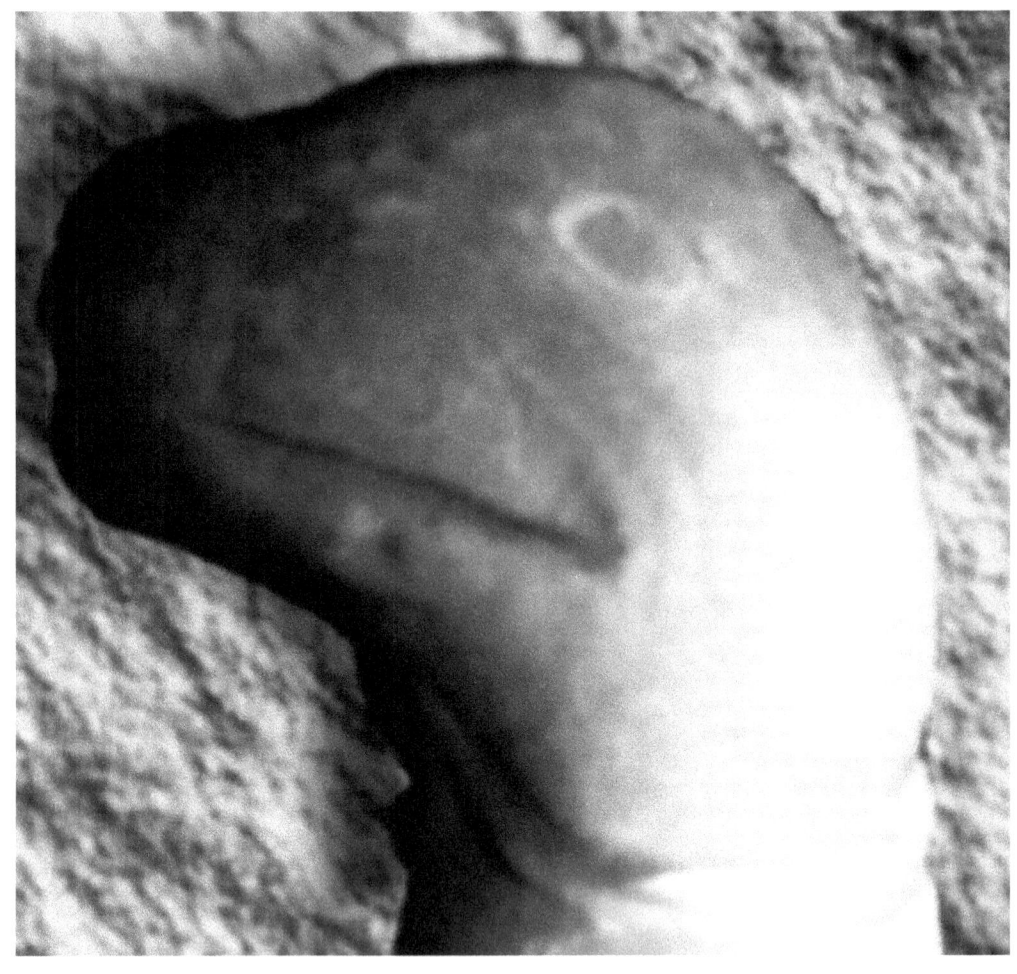

The head of the Shony? A stone image found at St. Peter's Church in Monkwearmouth, Sunderland, identified as a seal by some experts.

OPPOSITE:
Degraded but still visible carvings of dragons in the stonework at St. Peter's Church, Monkwearmouth.

Degraded but still visible carvings of dragons in the stonework at St. Peter's Church, Monkwearmouth.

Trow Rocks monster

I have come up with some strange things when delving into the Gazette's files but never a sea monster before!

I came upon the record of what appears to be the one and only sighting of a Tyne version of the Loch Ness monster quite by accident.

It happened 40 years ago when the crew of a salvage steamer, working on the wreck of the Eugenia Chandris, off Trow Rocks, South Shields, spotted the head and neck of a creature projecting five to six feet above the water.

Its back was eight or nine feet across and the 'monster,' which was black in colour, stayed in sight of the steamer the Black Eagle's pursuing motor boat for about ten minutes before disappearing below the surface.

Unfortunately, the crew couldn't get a sidways view of the creature to tell if it had the same distinctive profile as 'Nessie' — but isn't it fascinating?

SEABURN SENSATION.

BATHER'S ALARMING EXPERIENCE.

Considerable sensation was created among some of the bathers at Seaburn this morning. A young gentleman who is a regualr bather at the place and swims well returned from the water with a story that set all who heard it wondering.

He said that when he was about thirty yards from the shore, and was just about to come to the beach again, his right arm was struck a heavy blow by some floating body that must have been going through the water very fast, as the blow quite paralysed his arm for a few seconds. He had visions of sharks and other ravenous denizens of the deep, and swam rapidly back to the shore without seeing what had struck him. The upper part of his arm was bruised and bore a slight wound, and it was quite apparent that his nerves had been severely shaken by his experience.

What had struck him? Was it a huugry sea monster, or a harmless piece of water-logged wood

PRESS REPORTS

Above: An article about the Shony which appeared in The Sunderland Echo on August 17, 1906.

Left: An article about the Shony which appeared in The Shields Gazette on January 9, 1986.

Over: An article concerning a sea monster, which appeared in the Western Daily News on September 14, 1907. Some speculated that the creature, which was spotted off Gulla Stern Cove, Tintagel, was the same beast that had appeared near Seaburn just over a year earlier.

"THE SEA SERPENT."

Sir,—I write to tell you that on Thursday morning last at 11.45, I was sitting with a friend, Miss Hawley, on the N.E. cliff of Gulla Stem Cove, Tintagel, when we distinctly saw the black object which Mr. Dodgson has described in your columns to-day as a sea serpent. His description exactly tallies with what we saw, but his words hardly give the impression of the extreme rapidity with which the beast moved through the water, which was very calm at the time. A. C. MASON.

Eagle Cottage, Tintagel, September 14th.

Sir,—Although I do not accept the supposition that the great sea serpent exists, that which is so frequently mistaken for the beast must be capable of some intelligent and rational explanation. The observation reported by Messrs. Dodgson and Davis from Tintagel leaves the identity of the animal a mere conjecture, but, at the same time, points to its being one of the ribbon fishes (Regalecus).

This fish is known as the oar fish, or king of the herrings, and some twenty-five specimens have been taken at various times on British coasts. It may attain a length of more than twenty feet, and derives its name from the fact that it is laterally compressed like a broad band of ribbon. It is of a silver grey colour, with a few irregular black marks on the body. In the region of the head the dorsal fin is elongated, and forms a sort of crest. The fish is delicate in structure, quite inoffensive, and an inhabitant of deep water, coming only rarely to the surface, probably by accident or disease. It is not at all unlikely that this particular animal may, in the course of a few days, be found dead or dying washed up at some point of the adjacent coast, so that there still remains a possibility of some enthusiast clearing up this interesting point.

T. V. HODGSON.

Museum and Art Gallery, Beaumont Park, Plymouth.

Tynemouth Priory on the north side of the River Tyne. Some stories relate that tunnels under the priory were once inhabited by a sea monster.

Marsden Rock with members of The Society of Friends singing hymns and reciting poetry on the top. God knows why.

dog-eared cutting from the *Gazette*. To my amazement it detailed an encounter with the Shony from as recently as 1946.

In the summer of that year dredging work was taking place off the coast of South Shields. A ship named the *Eugenia Chandris* had been wrecked there earlier, and another vessel, a steamer named the *Black Eagle*, was heavily involved in the operation to break her up for scrap.

At some point, several crewmembers from the *Black Eagle* spotted a huge "head and neck" rising out of the water to a height of six feet or so. Seizing the moment, the sailors decided to give chase. Fortunately a motorboat was tied up to the steamer, and within minutes several crewmembers were making good speed towards the creature.

The leviathan appears to have seen the motorboat coming, and sped through the sea to a safe distance. Then, just as the launch almost caught up with it a second time, the creature sped back again. This cat-and-mouse game continued for quite some time, the Shony always staying tantalisingly ahead of its pursuers, who, according to the account, were trying to see the profile of the creature to compare it with descriptions of *Nessie*. To my knowledge they never did, for the Shony eventually tired of its game and then disappeared under the waves.

It is curious that the Shony in the 1946 report was reported as teasing its pursuers and leading them a merry dance back and forth along the coast, for the Shony has been linked with another creature from Scotland named the Shaillycot. The Shaillycot, unlike the Shony, is reputed to inhabit freshwater streams, but was also known then for its ability to imitate the sound and appearance of a drowning man. When would-be rescuers were just about at the scene, the Shaillycot would disappear under the waves laughing hysterically. It is interesting that both the Shony and the Shaillycot seem to have shared this incredible talent for teasing their observers.

On August 11, 1998 my wife and I were travelling with my father towards Whitburn along the coast road when we both happened to gaze out to sea simultaneously. Thirty yards or so from the shore we saw what I can only describe as a large, brown hump breaking the surface of the water, which was quite calm. I shouted, *"Can you see that? What on earth is it?"* (Or perhaps on reflection it was "What the ^&*$ is that?" Choose whichever version suits your sensibilities.)

My father was still trying desperately to focus on whatever it was we were looking at. Meanwhile, Jackie and I both watched in stunned silence as this large "hump" slowly sunk below the waves.

When we arrived home about 45 minutes later, the *Shields Gazette* was there waiting for us. On the front page was an article, which seemed to put the mystery to bed. Several people had spotted a bottle-nosed dolphin swimming near the coast, and the creature had indeed been immortalised on celluloid by one of the paper's photographers. Perhaps we hadn't seen the Shony after all, then. Maybe it was just Flipper.

But then the telephone rang, and I took a call from a local councillor who is a good friend of mine. His politics are crap, but he's a nice guy. He'd been standing in the queue in a fish and chip shop that lunchtime, and in front of him were two rather inebriated day-trippers who were arguing about the very same dolphin highlighted in the *Gazette*.

"*No way was that a dolphin wot I saw*", said Inebriate Number One. "*Wot I saw could have swallowed a dolphin in one gulp.*"

"*It must have been the dolphin, mate*", said Inebriate Number Two. "*I mean, stanstereasundunnit?*"

Interesting.
And then another *WraithScape* reader, Ivor Muncey, informed me that other sightings had been made of the creature off Trow Rocks in South Shields as far back as 1935. It was at this point I began to wonder whether there might have been more to the Shony than mere flights of fancy.

The Shony would not be difficult to recognise if you bumped into him in Marks and Sparks. He is very big, very long and covered in brown, oily-looking skin. He sports a large, fin-like protuberance, which runs the length of his spine and, in keeping with all good sea monsters, a peculiar mane of hair similar to that of a horse. He also has a set of razor-sharp teeth, which - according to quite a few Vikings - are capable of tenderising human flesh quite efficiently. Probably best to keep your distance from the teeth, then.

From time to time, those who reside along the coastal areas of the North East of England report seeing something large and vaguely threatening in the sea. Such sightings are usually dismissed as the product of an over-fertile imagination. In South Tyneside the creature has been labelled "Plessie" on the presumption that it may be a plesiosaur. Personally I don't give two hoots about the creature's taxonomy; I'm more concerned about those bloody teeth and their proximity to my buttocks.

I won't end this chapter by saying something sensationalist and dramatic like "The Shony is back", for the truth is I don't think it ever really went away.

Chapter Eleven

Tyneside's Fee-Jee Mermaids

Picture the scene. A schooner with white, billowing sails plunges through the choppy waters of the North Sea. The bearded captain, concern etched on his weather-beaten face, looks once again at what seems to be a small island in the distance. He raises his ornate, brass scope to his eye and gazes intently. There is no doubt about it. A beautiful woman is trapped on some rocks near this barren island, and she is frantically waving for help.

The captain is puzzled. What is she doing there? No time to think about that now, however. A boat would have to be lowered in an effort to rescue her.

Once again the captain gazes at the woman through his glass. She is beautiful, and has long, blonde hair, which billows in the wind. Strangely, she appears to be naked.

Suddenly the schooner shudders and the awful screech of breaking timbers fills the captain's ears. The ship has hit rocks hidden below the surface, and is already sinking into the water. The crew, panic-stricken, make for the boats.

The captain, still enchanted by the lady on the rock, sees her wave one more time before she dives into the water. To his astonishment he observes that she does not possess legs, but instead a tail. Too late, he realises what has happened. This was no trapped lady, but a mermaid – and she has lured the schooner and all aboard to a watery grave. With one last flip of her scaly tail she disappears into the depths, soon to go in search of other unsuspecting mariners whom she will lure to their deaths with her enchanting looks.

Well, when you've seen one mermaid you've seen 'em all, that's what I say. And can you imagine marrying one? She'd probably be running – sorry, swimming – off to the doctor's every five minutes suffering from fin-rot. And then there'd be the constant temptation to roll her in breadcrumbs, pop her in the deep-fat fryer and serve her up with chips and mushy peas. Seriously though, a huge number of sober, hard-bitten, no-nonsense seamen claim to have witnessed mermaids over the years. The idea that female, human-fish hybrids could be swimming around our oceans – like some bizarre cross between a cat-walk model and a haddock – seems *too* ridiculous to take seriously. So what is the answer?

Well, the idea that a race of half-fish, half-human hybrids once existed goes back a long way. The ancient Babylonians – always suckers for a good yarn – believed that much of their knowledge came from a race of fish-like gods who walked out of the sea one day and educated we land-lubbers in the arts of civilisation.

Mind you, I have always believed I got my education at the now-demolished Jarrow Central School, and I can't say that I'm overjoyed at the idea that I may actually owe my intellectual prowess to a tuna or a whitebait.

Other civilisations, including the Amerindians, the Chinese and the Celts also spoke of these water-dwelling weirdos. Even today, the Dogon – an African tribe that dwells in the land-locked country of Mali – say that in ancient times they were visited by fish-like space-beings called the Nommo who taught them, amongst other things, the science of astronomy.

Interestingly, the Dogon have known for hundreds – if not thousands – of years that the bright star Sirius has a small, dense companion that is invisible to the naked eye. They also have an intricate knowledge of the orbits of these stars – a knowledge that has recently been confirmed as accurate by modern science.

This is all fascinating stuff, of course, but it is hardly likely to help the poor fisherman who, whilst trying to catch his breakfast off the South Shields coast, is suddenly confronted by a buxom maiden who bears an uncanny resemblance to a goldfish below the bellybutton. Perhaps there is a more mundane answer.

Two aquatic mammals – the manatee and the dugong – have often been mistaken for humans. Up close you'd be hard pushed to understand why, but at a distance they can look quite human-like. It is certain that many mermaid reports have really been sightings of these two creatures. But this doesn't explain all cases. For one thing, both species are exclusively tropical animals.

In *The Times* of 8 September 1809, a Scottish schoolmaster named William Munro wrote that he had encountered a mermaid whilst walking along the coast at Sandside Bay.

If Munro was telling the truth - and those who knew him say he couldn't lie to save his life –

A Fee-Jee Mermaid in a display case
BELOW: Barnum's Fee-jee mermaid

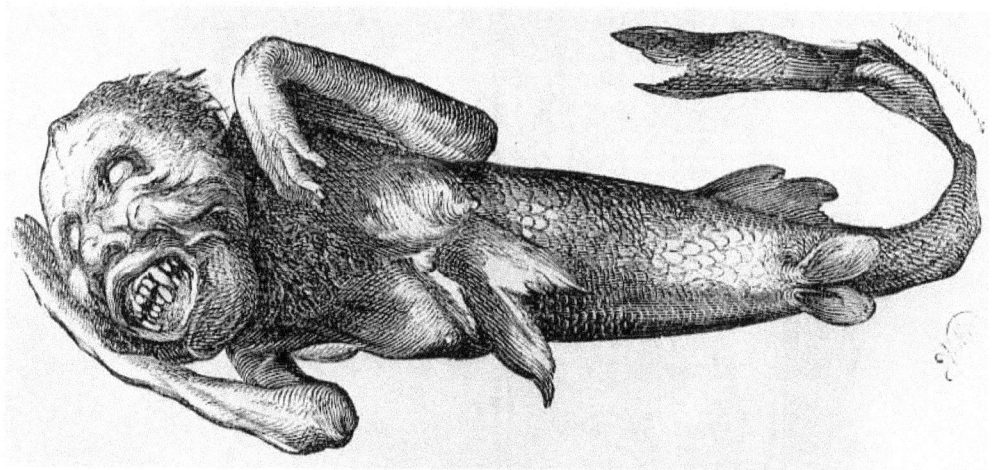

OVER: Misleading 19th Century advertising for fee-jee mermaids

EGYPTIAN MUMMIES,

and ancient Sarcophagi, 3000 Years old ; and an entire

Family of Peruvian Mummies;

the DUCK-BILLED PLATYPUS, the connecting link between the BIRD and BEAST, being evidently half each ;—the curious half-fish, half-human

FEJEE MERMAID,

which was exhibited in most of the principal cities of America, in the years 1840, '41, and '42, to the wonder and astonishment of thousands of naturalists and other scientific persons, whose previous doubts of the existence of such an astonishing creation were entirely removed;

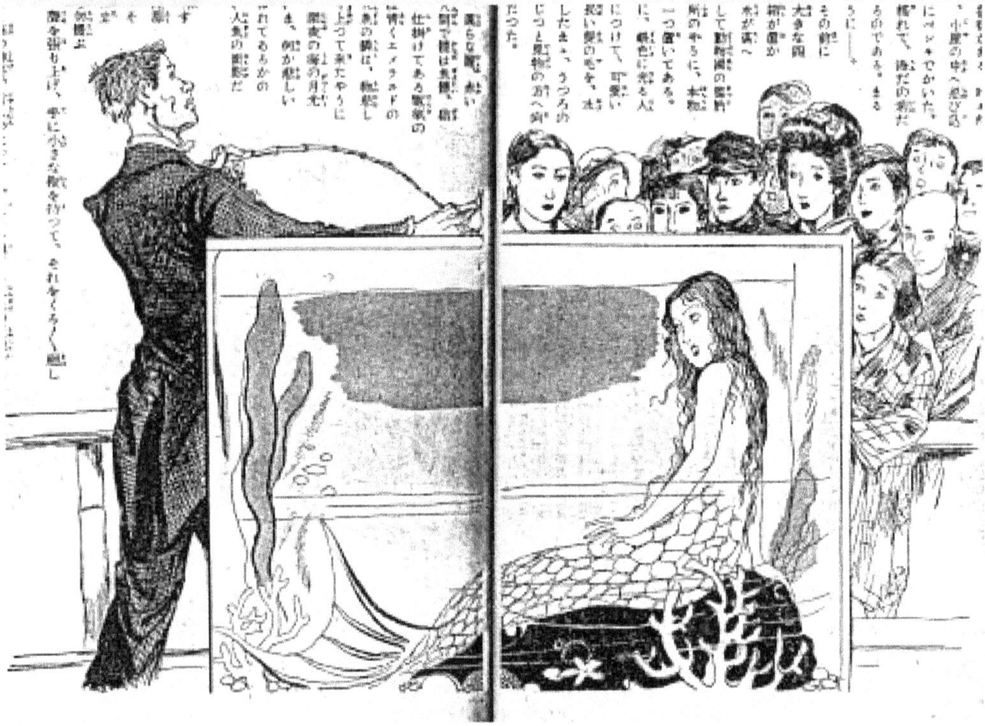

then what he encountered was certainly no dugong or manatee.

According to Munro the creature had flowing brown hair, ruddy cheeks, blue eyes and a mouth "of natural form". She had perfectly normal "breasts, abdomen, arms and fingers" too, and was actually combing her hair.

After approximately four minutes she plopped back into the briny and Munro presumably went in search of a stiff whisky.

As readers of this volume now know, there is some superficial evidence that the North Sea may indeed be home to a prehistoric sea – beastie; but naked ladies with fishtails frolicking around the coast? I've never seen such a thing. Still, I live in hope.

Some years ago, whilst ploughing through the *Shields Gazette's* photographic archives, I stumbled upon a photograph of a mermaid. What? A vision of loveliness combing her long, blonde hair as she basked in the sun? An aquatic enchantress sitting on a rock at Frenchman's Bay, waving to passing sailors?

Er, not quite. This mermaid – or it could be a merman for all I can tell – is hardly what the national tabloid editors would call "page three material". Its lower half looks like the tail end of a fish alright, but the top half comes straight out of a Hollywood "B" movie. It is not a pretty sight.

The only information attached to the photograph was that the creature – or rather the preserved remains of it – was the property of a lady who lived in Mowbray Road, South Shields. I had no idea when it was taken.

What I *was* able to reveal to *WraithScape* readers was exactly what this supposed mer-person really was.

Contrary to popular opinion, belief in the existence of mermaids is not entirely extinct. Sightings of the creatures continue to trickle in, several of which I have dealt with in columns and articles over the years. During the 19th century, in fact, the majority of people were prepared to acknowledge that such bizarre entities may have indeed existed.

Of course, Britain was a great seafaring nation during the same period, and some opportunists in foreign climes were more than prepared to cash in on our gullibility.

The animal in the photograph is typical of the fake "mer-creatures" which sailors would purchase for a small fortune in places such as Fiji, China and India. They would then present these trophies to their families, no doubt telling them how they braved storm and tempest to catch the thing after a fierce struggle near the Ryukyu Islands, or wherever.

In fact, taxidermists in the East skilfully created these mermaids in workshops. They were made by joining the tail of a fish, such as a sea bass, to the upper torso of a small monkey.

Then, having been placed against a scenic backdrop in an ornate glass case, they were ready for the local market. As Fiji was a hive of industry in the production of these bogus creatures, they came to be nicknamed FeeJee Mermaids.

A fake merman was actually displayed by Phineas T. Barnum when he toured Britain with his curiosity show in 1842. A good number of others are still on display in museums and in private collections up and down the country.

Within two weeks I was deluged with mail relating to the mermaid/merman photograph. Of course, I was absolutely delighted with the response from readers. (Two even informed me about another "mermaid" on display in a pub in Newcastle upon Tyne.) In fact, no less than forty e-mails arrived within the space of several hours directing me to a local gent's hairdressers in South Shields, where the mermaid had been, until recently, on display.

The hairdresser has, I believe, passed on, but his daughter told me at the time: *"The mermaid was allegedly caught off the coast of Japan 120 years ago. It was given to the family by a sailor who claimed it lived for three days after being caught."*

Nick-named "Sally" by the crew, the beast apparently survived for three days in a tank of water and then died. By all accounts it screamed hideously, chilling the sailors who watched over it to the bone.

Now being an old cynic, I did say that "Sally" was actually a fake creature made by joining the top half of a monkey to the bottom half of a fish. In all likelihood the purchaser was a sailor who picked Sally up in a Far Eastern market and invented the story as he sailed home.

But then again, who knows.

Actually, Tyneside sailors were also often fooled by a second type of fake mermaid. These are generally known as Jenny Hanivers, or, more uncommonly, Jenny Minivers. The use of this nomenclature is almost as interesting as the existence of the bogus mer-persons themselves.

A "Jenny Haniver" is a fish – almost always a ray, cuttlefish, ratfish, guitar fish or skate - which has been cut, shaped and otherwise altered before being dried so that it becomes rigid and retains its modified appearance. The desired result was achieved if the modified fish could be made to look like a miniature person; an angel, an extremely small child, or, more commonly, a devil, dragon or mermaid.

To produce a typical Jenny Haniver, a loop of cord was placed around the neck area and tightened. This resulted in the upper part taking on the appearance – very loosely – of a human head. The wings of flesh were then distorted either by rolling or stretching, depending on exactly what sort of "creature" the manufacturer wanted to create. The drying process would accentuate certain features of the fish's anatomy and create small holes that looked like nostrils and eyes. These could then be embellished to give the "creature" an even more human appearance.

After drying on a wooden frame, the finished product was then painted or varnished to preserve it.

Whereas FeeJee Mermaids were almost exclusively produced in the East, the most productive source of Jenny Hanivers was actually the docks at Antwerp situated on the estuarium of the River Scheldt.

The production of Jenny Hanivers at Antwerp went on for centuries, and there was a thriving trade in the selling of these "creatures" to foreign sailors. Just like the merchants of FeeJee Mermaids, sellers of Jenny Hanivers would regale sailors with stories of how the creature had been caught, how the crew had struggled to keep the thing alive to no avail, how it had screeched in the most unearthly manner as it went through its death throes... Amazing! Give me half a dozen.

As fast as the Belgian manufacturers of Jenny Hanivers churned out their mermaids on the production line, sailors from Tyneside would snap them up and take them back to Newcastle. Oh, they would pass on the odd one or two to relatives, sure, but the majority were sold to carefully selected customers who were believed to possess more money than sense. Hence, the north east of England became flooded with miniature mermaids, all of which had been caught by your sea-going uncle (under the most heroic of circumstances, of course) whilst fishing off the north east coast.

The earliest known expose of the Jenny Haniver hoax appeared in Konrad Gessner's epic *Historia Animalium* published in 1558. Gessner, born and educated in Zurich, was a Swiss naturalist who likely gained his interest in the animal world due to the fact that his father was a furrier. Gessner stated, quite accurately, that these so-called mer-creatures were actually modified fish and not, as they were often purported to be, mermaids and baby dragons.

The name Jenny Haniver is intriguing, as no one can say with any certainty exactly where or how it originated. One persuasive theory is that the name is a bastardisation of the phrase "jeune d'Anvers (Anvers being the French name for Antwerp). If this is true, the Jenny Haniver simply means "young girl of Antwerp".

Jenny Hanivers and even FeeJee Mermaids can still be seen occasionally in curiosity shops in Tyneside. Two pubs in Newcastle, to my knowledge, still have fake mermaids on show in the bar; one a FeeJee Mermaid and the other a rather battered Jenny Haniver.

Chapter Twelve

The Mouse Shrew

The 'Mouse Shrew' was at one time one of the most feared of cryptozoological animals in certain parts of the north east of England. Whether it actually existed as a fortean species in its own right is still a matter of debate, but one cannot dispose of the possibility in cavalier fashion.

Firstly, let's state the obvious; there are mice living in the countryside of the British Isles. Less well known, but quite prevalent, are shrews. There are five types of shrew in the UK, essentially divided into two separate groups. These are the red-toothed and white-toothed shrews. The red-toothed group has, as its name suggests, red teeth; or at least the tips of the teeth are red due to high deposits of iron in the enamel, which give them added strength.

White-toothed shrews are not at all common on the mainland. Lesser white-toothed shrews are usually only found on the islands of Jersey, Scilly, and Sark, whilst the Greater White toothed Shrew inhabits Guernsey, Alderney and Herm. [1]

There are three species of red-toothed shrew found on mainland Britain. These are the black and white water shrew, the grey-brown common shrew and the pigmy shrew or mouse shrew.

The mouse shrew (*Sorex minutus*) also known as the pygmy shrew, is quite widespread throughout mainland UK, but its habits mean that it isn't often seen. The mouse shrew is a diminutive creature and its body length is gen-

---

1. This species was discovered in Ireland for the first time in April 2008, during the preparation for this book.

erally only between 4 and 6cm. Its appearance is actually quite cute. It has a pointed, disproportionately long nose and a long, hairy tail. Its fur is grey-brown in colour. Although mouse shrews usually make themselves busy both above and below ground during night and day, their liking for holes, fissures and natural apertures in grass and other vegetation means that they are seen far less than would otherwise be the case.

The mouse shrew has a voracious appetite and needs to eat regularly, feeding on spiders, wood lice and other insects. Shrews are natural predators, and therefore have highly-developed defence mechanisms for when they find themselves in trouble from aggressors such as owls. They have an acute sense of smell, excellent hearing and a good sense of touch. They also have powerful stink glands, which make them unpalatable even after they are dead.

Mouse shrews are very territorial and will defend their "turf" aggressively. Their territory may be quite large – up to 2,500 square metres – but they will only leave it when looking for a mate.

Typically the mouse shrew will breed between April and October and will die the following November or December as their lifespan is a little over one year.

Despite its small size, the mouse shrew is highly aggressive. It will have nothing to do with other rodents despite its similarity in appearance. When the mouse shrew identifies its prey it will chase it relentlessly and doesn't seem to know the meaning of the word fear. This may explain why, in many parts of the UK including Tyneside and Northumbria, it came to be both feared and revered even by humans.

In legend, the mouse shrew is believed to be both harmless and highly dangerous at the same time. This needs some explanation. Anyone looking at a mouse shrew would immediately recognise that its small size would disable it from carrying out a serious physical attack upon a human being. No one ever got eaten alive by a mouse shrew or had their leg snapped off by one in a tussle. However, it's aggressive nature means that it could, if it felt so disposed, give you a sharp bite. This wouldn't kill you, of course, although their remained the risk of contracting an infection of sorts which, theoretically, could. So, why was the mouse shrew still believed to be dangerous?

In times past mouse shrews were believed to be highly poisonous. This belief may have sprung from the fact that domestic cats – one of the shrew's greatest enemies – would not eat shrews after killing them. As mentioned earlier, this was because the stink glands on the shrew contained a bitter substance that made them extremely unpalatable. When humans observed that the shrew's predators would not devour their catch, they may have assumed that the creature contained a poison of sorts.

Of course, it wouldn't have been long before someone put two and two together and made six; specifically, blaming the dreaded mouse shrew for poisoning a cow, goat or other sick animal. Whatever the truth, the negative reputation of the creature eventually grew beyond all sense and reason. Like so many other cryptozoological animals from the northeast region, the per-

ception took on a life of its own and became another creature; a spectral, darkly psychical animal that far outweighed its flesh-and-blood counterpart in every negative department.

In Northumberland, it was believed that if the mouse shrew crept over an animal, perhaps while it was resting, the unsuspecting beast would be, "bated [sic] with cruelty and intimidated with its limbs becoming of no value". It seems that farmers were not given to thinking that mouse shrew attacks were a rare occurrence. They were, according to folklorist William Brockie, repeatedly threatened with attack.[1]

Fortunately, there were remedies at hand for such emergencies. The standard method was to have an ash tree growing nearby. The standard procedure was to drill a deep hole into the trunk of the tree and place inside it a living mouse shrew. The hole would then be sealed with resin and/or wax and, eventually, the creature would die of hunger or lack of oxygen. As soon as the shrew had been sealed into the tree – even before it died, if necessary – a branch of the ash could then be snapped off, wrapped in a "red thread" and used to treat the afflicted animal. The switch was to be drawn across the ill beast's back several times (the exact number does not seem to have been important) and then it would miraculously recover as the "poison" of the shrew would be instantly dispelled from its system. In some traditions it was believed that the cure would only be affected after the encapsulated shrew had died. If an ash tree was not available, then a witch-hazel, witch-elm or rowan tree could act as a substitute.

Today, many of the old fears regarding mouse shrews have completely died out, but the legend lives on.

---

1. Brockie, William; *Legends and Superstitions of the County of Durham* (Sunderland, 1886).

Chapter Thirteen

Wandering Willie

"*Relatively few people in South Tyneside will now even be able to recall who Wandering Willie was, and certainly no one alive today ever saw him*".

These words were the opening lines of a column I penned for *The Shields Gazette* in November 2007, and they prove the adage that it certainly pays one to keep one's big, fat mouth shut when one doesn't know what one is talking about.

In one sense 'Wandering Willie' can be said not to be a "mystery animal" at all. We know who he was, where he came from and where he eventually ended up. However, there is an enigma attached to his life that, I think, merits his inclusion in this volume alongside the other creatures that we have devoted some considerable column inches to.

Wandering Willie was a celebrity in South Shields, but he died well over one hundred years ago. For those who don't know the story, let me introduce you to it.

In the summer of 1873, a shepherd from the Cheviots was tasked with taking a large flock of sheep to the Cleveland Hills. In those days, of course, you couldn't simply load them on to a livestock transporter – you had to walk, and the Cleveland Hills were a long distance to the south.

The shepherd was accompanied by a dog whose sole purpose was to herd the sheep, and at some point the shepherd, the dog and their large flock arrived at the ferry landing at North

Shields. They caught the ferry across to the other side, and made their up the bank towards a place where the flock could rest. Then, disaster struck. Something "spooked" the sheep and they bolted in a multitude of different directions. Before the shepherd could do anything to rectify the situation the surrounding alleys were filled with dozens of woolly intruders. Bedlam was the order of the day.

The dog was dispatched to round them up, but his task seemed hopeless. Nevertheless, over the course of several hours he managed to return the would-be escapees back to the ferry landing. The shepherd did a quick head count and announced that one was still missing. The dog was sent to look for it.

Another hour or two passed, and neither the dog nor the sheep returned. The shepherd did another head count and discovered that no sheep were missing at all; he'd simply miscalculated. Pressed for time, he resumed his journey and presumed that the dog would follow on later.

Alas, it never did. The dog was puzzled at his master's absence and decided to wait by the ferry landing for his return. For several months he waited faithfully, surviving on scraps fed to him by kindly passers-by. Some took him home, but the dog – now named 'Wandering Willie' by locals – would immediately decamp and return to the last place he saw the old drover.

Willie became something of a celebrity, and according to contemporary accounts quite conceited. He fell in with a local gang of young ruffians and engaged in all manner of mischief, particularly at the ferry and at the well-known Comical Corner. So bad did his behaviour become he was permanently banned from the ferry, which he liked to travel upon. Willie went from being famous to infamous in a very short space of time.

For no less than fifteen years Wandering Willie lived by the ferry landing, but eventually blindness and infirmity overtook him and he became incapable of fending for himself. A kindly ferryman eventually took him in and cared for the canine adventurer until, in the autumn of 1890, he passed away.

Wandering Willie's exploits lived on, though, and schoolchildren from all over the Borough were told cautionary tales of his derring-do activities. *The Shields Gazette* noted at the time that, *"It is sad to know that he went to his grave without ever again looking upon the face he loved so well"*.

And it's true. Wandering Willie never did see the old shepherd again, but it seems that he at least lived out the last years of his life with a degree of excitement and adventure.

In November 2007, I related to readers of my *WraithScape* column in *The Shields Gazette* the amazing tale of Wandering Willie, the sheepdog who became separated from his master and ended up spending the rest of his days in South Shields.

As well over a century had passed since the legend of Wandering Willie - as the locals called him – began, I doubted that anyone would recall the story. I was wrong. *WraithScape* reader

An original picture of Wandering Willie before his death. BELOW: Wandering Willie as he is now, on display in the *Turk's Head* public house in Tynemouth.

Comical Corner, where Wandering Willie caused mischief with a gang of local street urchins.

.Martin Embleton, whose father was personally acquainted with Wandering Willie.

Joseph William Embleton, who knew Wandering Willie as a child. BELOW: The Turk's Head public house in Tynemouth, where Wandering Willie is now on display.

The entrance to the old ferry landing at South Shields, where Wandering Willie lived. BELOW: The ferry landing at South Shields as it is today.

The now-demolished *Ferry Tavern*, where the ghost of Wandering Willie has been seen.

The Back of Fisherman's Wharf: a haunt of Wandering Willie.

Martin Embleton from South Shields contacted the *Gazette* and told me that his father actually knew Wandering Willie as a child.

Martin is a retired marine engineer. A sprightly 88 year-old, he also sang with his wife Norah and their efforts have been recorded onto a CD entitled *The Fantastic Sound of Martin & Norah Embleton.* Martin's father, also a marine engineer, was Joseph William Embleton. Co-incidentally, Joseph was born in the same year that Wandering Willie first arrived in South Shields and was quite familiar with him during his childhood.

Martin was born in 1919, and when he was eight his father began to relate to him the exploits of Wandering Willie and his daring adventures.

"Its true that Willie got into some scrapes, but he wasn't a bad dog. It was the youngsters he fell in with. He was well loved both in North Shields and South Shields, and that's how he got his name, Wandering Willie. He continually went between North Shields and South Shields on the ferry, wandering between the two towns".

Wandering Willie was really South Shields' equivalent of Black Bob the "Wonder Dog" featured in the *Dandy* comic many years ago, and he became something of a celebrity in his day. Joseph William Embleton lived in Alice Street, and told his son of how he would regularly see Wandering Willie by the ferry landing, still waiting for his master to return.

What happened to Wandering Willie after he died was something of a mystery, although Martin thought he had the answer:

"According to my father, who died in 1961, Willie's body was stuffed and mounted in a presentation case. Before he passed away he told me that it was still on display in a pub in North Shields, but I don't know which one".

Well, the author of this book decided to try and find out. He asked that if any *WraithScape* readers out there knew of Willie's whereabouts, would they please get in touch.

Martin said that he'd like to see a monument to Wandering Willie placed by the ferry landing:

"I think a plaque would be nice, bearing the true tale of Wandering Willie. It would be good for local people to read who don't know about it, and for tourists".

I agreed, but said that personally I'd like to see a full statue of Wandering Willie placed by the ferry landing, which would be a *real* landmark. I pleaded with any businesses in South Tyneside which would like to fund such a project to contact me. Wandering Willie was the Borough of South Tyneside's greatest canine character, and both Martin and I felt it was about time he got the recognition he deserved.

The response to my request for further information was phenomenal. Many readers contacted me and said that Wandering Willie had indeed been "stuffed" and was apparently on display in a public house in Tynemouth on the other side of the river. The problem was that only a handful could suggest which one. When I was just about to give in to frustration and abandon my search, Martin Embleton and a small number of other readers came to the rescue.

On Friday, 14 November, Martin contacted the *Gazette* and told chief reporter Angela Reed that he wanted to talk to me as he had some news. Angela left his telephone number on my answering machine, which I received when I arrived back from lecturing in London later that day.

It seemed that my columns on Wandering Willie had caused quite a stir, and an elderly neighbour of Martin had popped in to see him about the mysterious missing canine. He told Martin that the dog was on display in *The Turk's Head* in Tynemouth, not far from the castle and priory. Further, he kindly offered to travel across the Tyne that very day and photograph it. Alas, after arriving he found that his camera wasn't functioning properly and he was forced to return home *sans* pictures.

Not to be outdone, Mrs. H. and I went to *The Turk's Head* the following Monday with our daughter-in-law Rachel and our grandson Robin. As soon as I entered the bar I immediately saw Wandering Willie lying passively in a glass display case set into the wall.

The Turk's Head is an old Victorian beer bar that was built in 1850, and has a listed front facade decorated with white tiles. It provides a wide variety of entertainment for patrons, and also serves excellent lunches. Whilst Jackie and Rachel pored over the menu, and Robin chewed voraciously on a plastic rattle, I inspected the remains of the legendary Willie more closely. He was in fine fettle.

Time had not dishonoured old Willie. His coat was in perfect condition and in fact it was hard to believe he was dead. The taxidermist had done an excellent job. I asked the manager if I could take some photographs, and he kindly agreed.

But there are a number of puzzles about Wandering Willie that need to be addressed. I have in my possession several photographs of him that were taken when he was alive. It seemed to me – and one of the bar staff – that Willie's coat seemed considerably darker now than in the original photographs, particularly around the tail area. His nose also seemed a little longer in the pictures than it did on the preserved corpse. Later, when I examined the photographs, I concluded that the apparent discrepancies were simply tricks of light and shade. The two dogs were, I determined, one and the same.

Another discrepancy concerned the date of Willie's demise. A plaque in the pub said that he'd been preserved for posterity in the year 1880, whereas my research indicated that he'd died around a decade later. In truth it didn't matter; Wandering Willie was still here, in a sense, and the exact time of his demise was not an important factor.

Other correspondents told me tales that had been passed on to them by elderly relatives. Willie's ghost had apparently been seen in a number of locations, including the site of the old ferry landing and near the now-demolished public house called *The Ferry Tavern*. *The Ferry Tavern* was a pub that had seen great times and bad times, but most Sanddancers, as residents of the town are called, agree that our cultural heritage was diminished with its passing. The inn had also been haunted by the wraith of a young girl dressed in Victorian clothing who had seemingly died near the premises, but whether she had any connection with Wandering Willie I cannot say.

Both South Tynesiders and North Tynesiders lay claim to Wandering Willie as "their" dog. There is an old saying that possession is nine tenths of the law, and if its true then the good folk of Tynemouth inevitably have a better claim upon him. The residents of Tynemouth have his mortal remains, and the people of South Shields have fond memories; so again I would suggest that it matters not – Wandering Willie has enriched the lives of them all.

The Turk's Head is known colloquially as "The Stuffed Dog" because of Willie's presence, and no one seems to mind. Should his remains have been given a decent burial instead of being preserved and put on display? Probably, but that is another argument for another day. The 19th century was a different era and governed by moral norms different to those that currently influence our own thoughts and behaviour. At least we can say that his preservation has kept alive his memory and exploits, and that is not a bad thing.

Willie seemed to have an almost psychic rapport with the urchins he befriended, and this contributed immensely to the legendary status he would come to enjoy. Some of his personality traits seemed almost human; his mischievousness, occasional moodiness and, now and then, precocious behaviour. If his spirit still does occasionally visit his old haunts – if you excuse the pun – then we should be thankful.

Chapter Fourteen

The Monster Lobster of Trow Rocks

In the pages of this book I've detailed a goodly number of strange creatures which – allegedly, of course – inhabit the north east of England; giant rabbits, killer mice and hairy hominids to name but a few. However, perhaps the strangest beast of all is an aquatic monster known as the 'Giant Lobster of Trow Rocks'. Whether it ever existed or not – it may still be alive, for all I know – is something that at this juncture I cannot determine. I truly hope that it did once skitter around the bed of the North Sea, and there is just a glimmer of hope that it may still be out there. Mind you, I'm not so sure I'd ever like to meet it.

When I was researching the story of Wandering Willie, I was contacted by a reader of my *WraithScape* column called Alex Carrington. As we entered into dialogue with each other via e-mail, Alex almost casually mentioned the giant lobster, a creature that had hitherto escaped my attention. Indeed, I'd never heard of it until Alex dropped it, like a cryptozoological bombshell, into the conversation. I had to know more.

According to Alex, the giant lobster of Trow Rocks has its origins shrouded in mystery. Alex was unable to say where the story had originated or indeed how much currency it had ever gained amongst the local populace. Still, it is a fascinating tale indeed and needs to be recounted.

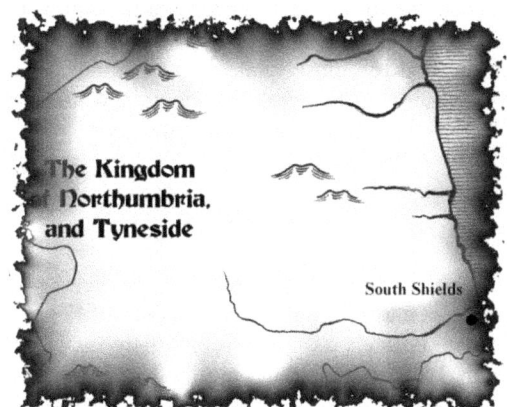

It seemed that in between the two World Wars, a colossal dock gate had been removed from a local shipyard and earmarked for transportation to another locale to be broken up and sold as scrap. With the help of the local history library at South Shields, I managed to put some meat on the metaphorical bones and uncover the details of the event.

On Sunday 19 October, 1919, the steel dock

gate, weighing many tons and which had previously been housed in the River Tyne, was hooked up to a flotilla of tugs and slowly drawn out to sea. As the procession passed Trow Rocks, where the currents are notoriously fickle, a storm broke out and the gate broke free. Within seconds it had plunged to the seabed, the top half sticking out of the waves in an ugly display of defiance against the tempest.

Shortly after the accident, the gate was purchased *in situ* by a Sunderland businessman called Potts - the father of the well-known Councillor George Potts. Potts the elder had hoped to have the gate broken up and sold, making a tidy profit into the bargain. However, the Trow Rocks tides again created insurmountable problems. Another difficulty was that the gargantuan frame had been filled with metal punchings just before transport in an effort to create ballast. The extra weight compounded the problem of moving the gate from the seabed even further, and the project was eventually abandoned.

In 1939 the gate was sold at a snip to the British Iron & Steel Company who also had plans to salvage it. Alas, the company's greatest efforts and most sophisticated equipment were no mach for the elements, and they to gave up any notion of cashing in on the man-made steel mountain lying just off the coast of South Shields.

Over the succeeding decades the huge gate rusted. Bits would drop off on a regular basis, and eventually only a few, small girders and other protuberances were visible above the waterline.

Fast forward the historical tape to the 1960s, and we find Alex Carrington as a child. With friends, Alex was accustomed to playing on the sands of South Shields beach. From the beach it was still possible to see some of the wreckage of the dock gate sticking out of the water. The twisted conglomeration of metal created a strange silhouette that led itself to fantastical interpretation by childhood minds, as yet unfettered by convention. An island inhabited by pirates, a half-sunken galleon filled with doubloons…? One could speculate forever, and the youngsters who gazed upon it certainly did.

It seems that, at some juncture, the story surrounding the wreckage took upon itself a rather darker aura. To Alex and friends, it was not so much the wreckage that was important, but what allegedly lived underneath it. Word started to spread that a giant lobster – or in some versions of the story, a giant crab – had taken up residence underneath the wreck of the gate and was extremely antagonistic towards those who intruded too closely into its new abode. The crack was that bathers or divers who swam too closely to the creature's domain would be snapped up by one of its giant claws and never seen again – or at least, not in one piece.

Alex summed up the whole affair wonderfully when he told me, *"We were all as mad as a box of frogs, but we had some imagination"*.

Absolutely true, of course; but I learnt many moons ago "never to say never" in the weird and wonderful world of forteana. If there is room on planet earth for a the 'Giant Rabbit of Felton' and the 'Laidley Worm of Spindleston Hough', then, unlikely as it may seem, there may also be enough living space for the 'Monster Lobster of Trow Rocks'. After all, the North Sea is a

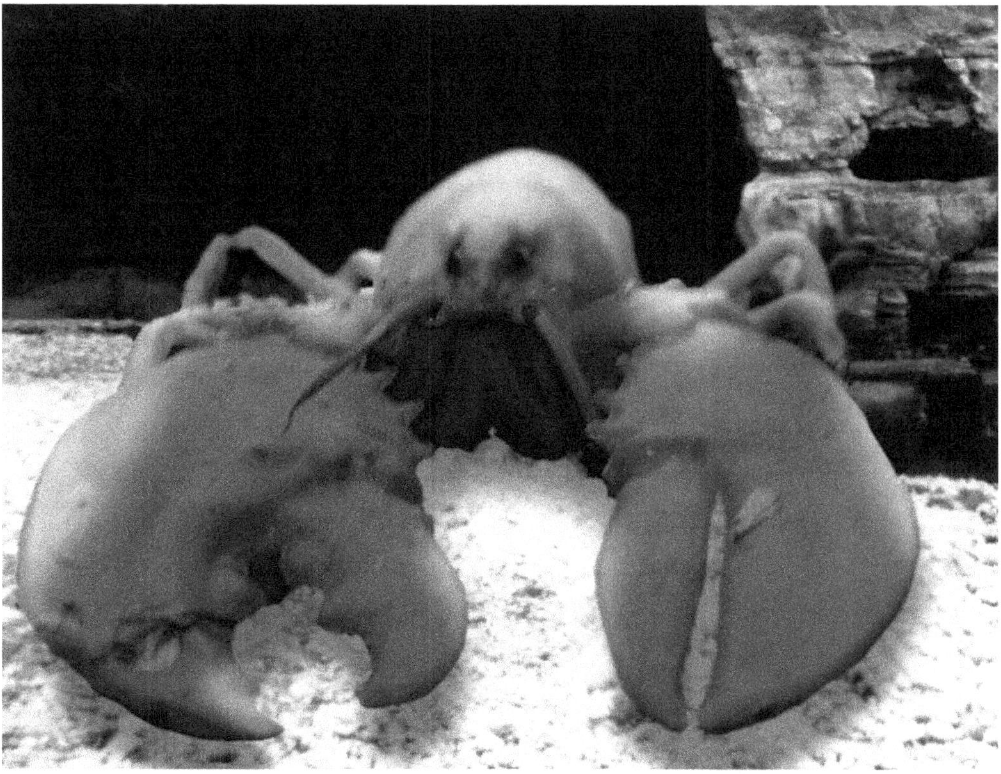

A giant lobster at Trow Rocks? (Artistic representation). BELOW: The huge dock gate just off Trow Rocks, allegedly the home of the Monster Lobster.

pretty big place.

Without delay I rang up some of my contacts in South Shields and asked them if they had heard of a colossal crustacean with a penchant for gobbling up unsuspecting tourists as they floated on the briny on their inflatable loungers. Alas, they hadn't, but they promised to ask around. With the luck of the Devil behind me, I hit paydirt. My correspondent – who doesn't wish to be named, lest he be labelled by his drinking companions as…well, as mad as a box of frogs – recalls that "around 1960 or 1961", a small article appeared in *The Shields Gazette* about a curious incident that took place at Trow Rocks one evening.

"I can't remember the details, as it's so long ago now, but to the best of my recollection it involved a student who was walking near the shore when he saw what he described as a 'huge monster' on the sand. He stared at it for more than a minute, during which time it remained perfectly still. This made him think that it might have been a statue or a model of some kind, but then it scuttled down the sand into the sea. That's all I can remember, really".

I haven't been able to locate the article in the *Gazette* archives, but I intend to keep looking. The question is whether it just possible that there may have been some factual basis to the sighting.

Ironically, the answer to this question may have been brewing in Germany during the very time I was researching the 'Wandering Willie' story which precipitated my contact with Alex Carrington.

At the latter end of 2007, a huge, fossilised claw was discovered in a quarry near Prum in Germany. The appendage was identified as having belonged to *Jaekelopterus rhenaniae*, a 400,000,000 (or thereabouts) year-old giant sea scorpion which, in its day, had very little to worry about in the way of natural predators. All things considered, it was top dog and terrorised the swampy marshes along the coastlines of its natural habitat. *Jaekelopterus rhenaniae* could grow up to nine feet in length at least, and maybe much larger. The finding of this large specimen told the scientific world that sea scorpions had been able to grow much, much larger than was previously thought. It was, in fact, approximately eighteen inches larger than any previously discovered *eurypterid*.

There has been some considerable debate about whether *Jaekelopterus rhenaniae* and its near relatives could have left the water and walked on dry land. Some experts say nay, whilst others have postulated that it may have left the water and walked upon the land quite frequently. The jury is still out, but we must at least consider it a possibility.

Intriguingly, the fossilised remains of giant water scorpions have been found in Scotland, so it seems almost certain that they also inhabited the northeast coastline off what is now South Shields. The sea scorpion doesn't look identical to a lobster, but to a slightly inebriated student seeing one after dusk the two might have been nigh on indistinguishable.

Of course, although creatures that resemble sea scorpions still inhabit our oceans they are all relatively diminutive when compared with their prehistoric counterparts. As far as we know, the giant varieties all died out millions of years ago. But, then, they said that about the coelacanth, didn't they?

It seems a racing certainty that the legendary monster lobster that terrorised the psyches of Alex Carrington and pals was generated by the alleged sighting of such a creature by the aforementioned student in the early 1960s. Of course, it is possible that there is simply no truth in the story at all, and that the entire episode is naught but folk-fiction. On the other hand, there must be a slim possibility that what the witness saw was a giant sea scorpion, the *specie* of which has survived, against all the odds, to our present day.

But there is a third possibility. Over the years I have written up a number of reports concerning prehistoric animals that have been seen in locations where common sense dictates they shouldn't – indeed couldn't have been.

In 2001 I was asked to travel down to Norwich and address the members of the Norwich Earth Mysteries Research Society on my experiences investigating the weird, the wonderful and the downright spine tingling. Never one to miss an opportunity to hold court – particularly when the meeting is to be held in an old inn – I immediately accepted.

For nearly an hour and a half I recited a litany of stories about ghosts, UFO sightings and sea monsters, only pausing to drink the glasses of Newcastle Brown Ale which they were continually forcing into my hand – honest.

Afterwards, over yet another glass of Brown Ale, a rather gregarious chap went on to tell me a very interesting tale indeed.

It seems that some years previously he went with some friends into the country. After nightfall they suddenly noticed that an eerie silence had descended over the area. Then, to their surprise, a gigantic bird-like creature floated over their heads. I say 'bird-like', as this certainly wasn't a sparrow out looking for a midnight snack or a budgie making a dash for freedom.

"It was about thirty feet up", he said *"and as soon as we saw it we knew exactly what it was – a fully-grown pterodactyl"*.

As soon as I returned home I wrote up the experience in my weekly column. Now before you start making sarky comments about this gentleman's alcohol intake – which on the evening in question was considerably less than my own - bear in mind that there were several witnesses to this event. It should also be remembered that, extraordinary though it may seem, reports of pterodactyls are often received in the USA, where some researchers think that it is just possible that a small colony of them may have survived the last few million years unnoticed.

Mind you, although I'd be the first to admit that the concept of a few pterodactyls soaring around in the remoter parts of the Nevada Desert or wherever is not impossible; the idea that they may actually have taken up residence near Norwich does prove problematic. Why haven't they been seen more frequently? My colleague Jonathan Downes, from the Centre for Fortean Zoology, believes – as do I – that what these people may have observed was not a "real" pterodactyl but rather what is known as a 'zooform' creature.

Zooform creatures are not flesh-and-blood animals, but spectral beasts, which bear a similarity to ghosts and apparitions. They can appear and disappear at will, and are never caught. Their origin and purpose is a mystery, but the number of people who have witnessed them is so great that their testimony cannot be ignored.

Maybe, then, this small group of startled onlookers were allowed a quick glimpse into the past, catching a scene from a legendary time when giant dinosaurs really did walk the earth – and soar through the heavens as well. It is also just possible that the young lad who witnessed the monster lobster at South Shields may also have seen a zooform creature; a seemingly real image of a giant sea scorpion from a distant era that was generated in some way we by a process we don't quite understand.

The Monster Lobster of Trow Rocks is an enigma that, in all probability, will never be solved. Mind you, if any film directors or producers out there would like to make a movie – maybe entitled *Monster Lobster vs. the Shony: The Final Conflict* - for a sum not unadjacent to £2,000,000, I'll be happy to co-operate...

Chapter Fifteen

The Dragon of Long Witton

Longwitton – sometimes called Long Witton, just to confuse tourists – is to be found in the heart of Northumberland. Visitors will be enchanted by the presence of trees, houses, motor cars and, more exotically, the odd dragon.

The tale of the Long Witton Dragon is a strange one, and is likely to terrorise young children if told to them as a bedtime story – an added bonus, I'm sure you'll agree.

There are some picturesque woods near Long Witton, and therein can be found three archaic wells, the waters of which are said to possess miraculous powers. As the locals found these powers quite handy, they would oft visit the wells and draw out bottles of the stuff for sundry purposes. One day, so the legend goes, a ploughboy was moved to visit the spot and collect some of the healing H^2O, although how exactly he was going to use it is something now buried in the sands of time. However, on his approach he was somewhat alarmed to see that a dragon had got there before him. Without so much as a flicker of embarrassment it was drinking from one of the wells by lapping at the water with its long, black tongue.

"The cheeky bastard!" thought the ploughboy, or at least, that's what I think he would have thought. The dragon, suddenly becoming aware of the youth's presence, promptly disappeared.

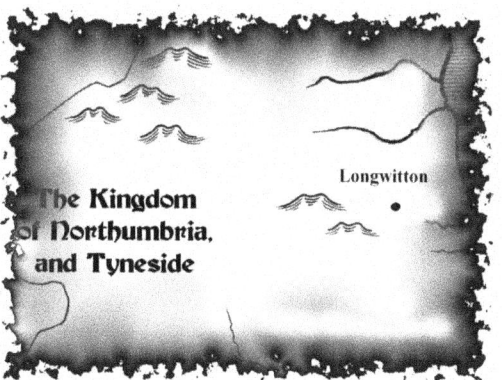

But it didn't go away. It just couldn't be seen, which is not the same thing. The ploughboy, though, was not to be fooled. He could hear its scales scraping together and couldn't help but notice that its stinking breath was still washing over him. Dragons can be bad, but dragons with halitosis are even worse. I know – I have a mate who married one back in the 60s.

From that time forth, there was a Mexican stand-off of sorts. The dragon wouldn't let the locals go anywhere near the wells, and the locals had no intention of going anywhere near the wells because there was a dragon there.

Now in all good dragon stories there is usually a knight who enters the vignette late in the day, but just in time to slay the dragon. The Longwitton tale is no different. At some juncture a young warrior – Guy, Earl of Warwick - heard about the beast that was troubling the good folk of the village and decided it would be a perfect opportunity for him to cut his teeth in the dragon-slaying business, which was all the rage in those days. Now it seems that the knight was a devious young cove, and had in his possession a magical salve that he'd picked up on his travels. By anointing one's eyes with it, one could see things which were otherwise invisible. Confident that a liberal application of *Invizisalve* (available now from all good stores) would do the trick, he headed off for the woods in search of the dragon. He found it, and immediately started hacking at it with his sword. *Hackity-hack* he went, all day long, but every time he cut the beast's flesh the wound would instantaneously heal up.

Bugger, he thought, before finally giving up and, in a state of physical exhaustion, riding home to Longwitton.

"Kill the dragon, then?" the villagers asked repeatedly.

"Er…no, not quite. I'm going back tomorrow to finish it off".

The following morning he once again applied the magic salve to his eyes and set off for the woods. On finding the dragon he engaged in a good bit of smiting – it went on for the whole day, in fact – but every time he cut it those bloody wounds just kept healing up. Worse, the thing just seemed to be getting stronger and stronger.

"Stuff this for a lark", he thought at sundown, and once again he headed back to the village. This was getting embarrassing. He'd have to finish it off the next day, otherwise his reputation for hacking and smiting was going to go right down the pan.

The following morning, at dawn, he went back to the woods and did a bit more smiting and hacking, but this time he took careful notice of the dragon's movements. He noticed that every time he backed away from the creature it seemed curiously reluctant to follow. In fact, it only seemed at ease when it was in close proximity to the wells full of miracle water. Then he noticed something else; the dragon's tail was actually draped over the edge of one of the wells and submerged in the water itself. Then, at last, the proverbial penny dropped. The reason that the dragon's wounds were healing up so fast was because it was constantly in contact with the miraculous healing powers of the wells and their contents. If he was going to beat *this* sly bastard he'd have to put his thinking cap on.

Eventually Guy came up with a plan. Drawing his sword, he charged at the beast with a blood-curdling yell and, yea and forsooth, did a bit more hacking and smiting. But then he suddenly stopped and pretended that the beast had mortally wounded him. He staggered away from the dragon, probably bellowing, "Oh, woe is me!" and a few other melodramatic one-liners just for effect. The dragon thought to itself, "Now I've got you, you cocky git", and chased after the knight as fast as its scaly legs would carry it.

But it was all a trick. The knight wasn't wounded at all. As soon as the dragon left the vicinity of the wells, our young hero circled behind it and cut off any chance of retreat. Now, the dragon was not only separated from the wells, but also from the healing powers of the water inside them. The knight renewed his smiting and hacking with gusto, and this time the dragon's wounds didn't heal up. The dragon bled to death, and the knight rode back to the village with a swagger.

"Kill the dragon, then?"

"Yeah, no worries. A piece of piss, mate, I'm telling you".

The next day the villagers held a huge barbecue outside (you should never hold them inside) in honour of the valiant young dragon slayer and the incredible efficacy of *Invizisalve* (buy now while stocks last).

Alright, we've had a bit laugh, but on a more serious note there are some curious features about the Longwitton tale that bear further scrutiny. For example, one of the three wells is known colloquially as The Eye Well, which may be because it makes your eyes well, although then again it may not. Regardless, it is said that if one applies the water from the well to diseased eyes, they will be healed. I can't help but wonder if there is a subtle connection between the powers of the water to – allegedly, of course – heal diseased eyes, and the "eye salve" used by Guy, Earl of Warwick. In both instances we have an allegedly miraculous substance being applied to the eyes in an effort to immediately improve one's circumstances. Maybe I'm looking for parallels where they aren't any, but it's an intriguing thought.

For a more detailed account of the Longwitton Dragon, I'd strongly suggest you read Richard Freeman's excellent *Dragons – More Than a Myth?* (CFZ Press, 2005). Richard treats the reader to a detailed description of the dragon's physical appearance and, in typical Freemanian style, a bloodcurdling account of the final battle between the beast and Sir Guy.

Roll up, roll up, laydeez anf jennelmayn...get yer magic eye salve here.....

The White Horse of Farding Lake

Essentially, the story of the white Horse of Farding Lake, South Shields, relates that "Elizabeth, the wife of Sir Hedworth Williamson", was out riding in the area one day - on a white mare, of course - when she suddenly disappeared without a trace. Extensive searches were carried out throughout the entire locale, but she was never seen or heard of again. In memory of his wife, Sir Hedworth duly commissioned a local man to paint a white horse on the cliff at Farding Lake near the old limestone quarry. Well, at least that's the story. However, there seems to be no evidence whatsoever that a "Lady Williamson" ever went missing at all, or that her death was in any way unnatural. In fact, all of the "Lady Williamsons" of this era – and there were a few – seem to have outlived their husbands. Even the date of the disappearance is debatable, although we can be certain that if it occurred at all it was in the latter years of the 19th century.

One version of the story relates that Sir Hedworth and Lady Williamson were seemingly in the habit of riding along the beach at Marsden, and regularly held picnics under the cliff at nearby Farding Lake; you know the sort – cucumber sandwiches, stuffed pheasant and champers – all packed into a wicker basket. One day Sir Hedworth had to travel away on business, and consequently Lady Williamson made an ill-fated decision to go riding on her own.

At some point, Elizabeth seems to have wandered from Farding Lake to the coast at Marsden and taken her horse down onto the sands. Eye-witnesses claimed that they had seen her galloping towards Marsden Rock. Then, apparently, she took the unusual step of riding into the large aperture which led into the interior of the Rock itself. Why she should do such a thing will probably remain a mystery for the rest of time. However, what we are also told is that although Lady Williamson *entered* the cave under Marsden Rock on her horse, extraordinarily, she never came out again.

Search parties looked long and hard, but found no trace of either Lady Williamson or her horse. In fact, for two, long years Sir Hedworth carried on looking for his missus. He held out no hope that she was still alive, but he loved her to bits and was desperate to give her a decent burial. Unable to give her a good send-off, he eventually died of a broken heart.

According to another version of the tale, before his death Sir Hedworth apparently lost all interest in his estate. Eventually he gave all his prize horses to his ostler, a man by the name of Wareham. Wareham appreciated this, and in memory of his employer's kindness apparently painted a white horse with both tar and whitewash on the small cliff at Farding Lake. This we know to be untrue, as the horse was painted on the cliff long before Sir Hedworth was even a twinkle in his mother's eye.

Another tale about Elizabeth's enigmatic absence concerns her disappearance inside Marsden Rock.

Seemingly, there used to be a cave under the rock which, in turn, led to another even deeper under the coastline. Apparently, accessing the deeper cave was difficult, but local smugglers had devised a way of entering it safely and had taken to storing their contraband there away from the prying eyes of the preventatives or Excise Men. Due to the tidal flow it was only possible access the cave at certain times, and a large slab of stone covered the entrance in such a way that it would be impossible to determine that anything but solid rock lay behind it unless you were told. One correspondent told me that, "The stone was balanced with another stone underneath it. If you moved the stone underneath then you could move the covering stone or 'seat' quite easily".

The covering stone apparently had a rope attached to it, which enabled it to be pulled back into position from underneath. Lady Williamson seemingly became trapped when those looking for her up above moved the stone and jammed it, thus sealing her in the subterranean tomb forever. The likelihood is that she drowned, or perhaps collapsed exhausted after wandering around the network of tunnels that lie underneath the bay. Her body was never found.

The greatest enigma, of course, is how on earth Elizabeth's horse was able to descend into the cave system with her. If it was difficult even for the smugglers to gain access, how did a large mare manage? Perhaps it didn't, and Lady Williamson went down into the cave alone whilst her horse remained up above. Indeed, one legend says that her horse died two years later and was buried at Farding Lake.

According to local folklore, Elizabeth Williamson was never found, although a funeral of sorts was held for her. Instead of a body, the coffin was weighted down with rocks wrapped in canvas. To prevent gossip, her family then spread the fictitious story that she had died of rheu-

The White Horse of Farding Lake – a cliff painting, the origins of which have been debated by researchers for decades. BELOW: The tomb of Lady Elizabeth Williamson, who allegedly disappeared under mysterious circumstances in the late 19th century. Her story is inextricably linked to the legend of the White Horse of Farding Lake.

The cottage at Farding Lake owned by Mr. Ness BELOW: Two books from the mid-19th Century that touch upon the mystery of Farding Lake

matic fever.

Local historians are generally of the opinion that the tale of Elizabeth Williamson's disappearance is a fiction, but are far more divided when it comes to establishing the truth about the white horse painted on the cliff. The truth is that there are simply dozens of stories concerning its origin, each one with a devoted band of followers. Debate over the matter has raged for years in the local press, and shows no signs of slowing down. In 1979, one local author told the *Sunderland Echo* that one story was essentially a cock-and-bull tale, whilst another was just a yarn that spread throughout the community but had little more credence to it. Janis Blower, who writes a local history column for *The Shields Gazette*, has commented on the legend on numerous occasions over the years, but has to my knowledge refrained from suggesting a definitive answer. No one seems to know the truth about the strange white horse painted on the cliff.

The fact is that the legend of the White Horse at Farding Lake goes back for centuries, and the truth behind the tale is far more complex than one would imagine. A detailed study on the mystery will appear in our forthcoming book, *The Order of the Dragon*, co-authored by Mike Hallowell, Jonathan Downes and Richard Freeman, to be published by the CFZ Press in 2009. However, there is another mystery about the white horse of Farding Lake, and this one has a far more cryptozoological twist to it.

To unravel the mystery of the white horse of Farding Lake, it is essential to break it down into its individual components. There is the historical mystery of who painted the horse on the cliff, the psychic-cum-spiritual mystery of the recurring white horse motif in local folklore and – just as importantly – the mystery of a ghostly white horse that has been reportedly seen on numerous occasions in the same area. It is this latter mystery which concerns us here.

There seems little doubt that the white horse enigma had its origins with a real, live, flesh-and-blood horse. From this one conventional animal sprung a plethora of myths and legends, including stories that, at certain times, it returns to walk the roads and byways around Marsden in ghostly form. Some say it has glowing red eyes, and that to touch it will bring about the severest of symptoms – namely, death. Although the white horse is said to be a "ghost", the reader should bear in mind that this is simply a colloquialism. Witnesses are definitely seeing *something*, but it has more in common with Black Dog manifestations than hauntings.

I've received two reports over the years, from readers of my *WraithScape* column in *The Shields Gazette*, who claim to have either seen the white horse at Farding Lake or know someone who has.

The first came from a chap who simply signed himself as "S. Cameron". Mr. Cameron claimed that on Wednesday 8 January, 1997, he had been visiting a friend who lived in Lake Avenue, South Shields. At approximately 11.50pm he left for home. Just as he was about to get in his car, and unfortunately after his friend had shut the door, he saw a large, white horse clip-clopping down the street. Not only was it unusual to see a horse out at that time of night, but the really peculiar thing was that it was riderless.

According to Mr. Cameron, *"There was something eerie about it. It seemed to glow...I can't really explain it. It honestly scared the life out of me. It stared at me as I got in the car, and I know this sounds daft, but it looked arrogant. I just got in my car and drove away"*.

The second story was given to me by Gazette reader Bob Cutters of Mortimer Road, South Shields. Bob telephoned *The Shields Gazette* and told a journalist that a friend of his had once seen the fabled white horse in 1963. He couldn't remember the exact date, but said that he'd been drinking in a local pub one afternoon and decided to pay a visit to his mother who lived in King George Road. On his way he remembered that he had to pass on a message to his friend who lived by the coast, so he decided to make a slight detour.

"I was driving up the coast road, and I remember that the weather was really atrocious. It was raining heavily. I was surprised to see my pal outside of his house in nothing but his shirt sleeves. At first I thought he must have been looking for his dog or something, but when I pulled up he jumped inside the car. He was really shaking. He said that he had gone outside to put some rubbish in the bin when he had a funny feeling that something was staring at him. He turned around, and was shocked to see this bloody great horse looking at him over the garden fence.

"According to my mate, he got an even bigger shock when the horse just disappeared. He went down the path and out of the gate to look for it, but it had just gone. Then he got the same feeling and turned around again. There it was behind him, just staring at him. What really scared him was its eyes. He reckoned they were big and glowing. Then it disappeared again, just before I pulled up in the car".

The white horse of Farding Lake lives on, and I for one can't wait for its next appearance...

Conclusion

When I took my first tentative steps upon the Fortean landscape, I harboured a naïve hope that, one day, I might come to understand things better. I think I even aspired – ridiculous, I know – to solving the odd mystery or two. After several years it dawned upon me that it was not to be. Fortean mysteries are not there to be solved; they are there to be revelled in. Seasoned researchers know that no enigma within the world of paranormal research is ever truly laid to rest. The gods allow us to *think* we've solved something for a maximum duration of five minutes – just enough time to shout "Eureka!" - before cruelly throwing a handful of inconvenient facts into the pot to spoil our great discovery. They must be having a whale of a time on Mount Olympus.

Ironically, having just about given up the hope of truly making any headway in anything, my life as a professional writer has become infinitely more enjoyable. Freed from the burden of solving things, I can now enjoy Fortean phenomena simply for what they are. I no longer study, I just gaze in wonder. Writing this volume has been the most enjoyable exercise I've engaged in, book-wise. Do I understand much more about the bizarre creatures which amble across our landscape when the lights are dimmed? Not really, but I am more convinced than ever that they are there – and certainly worth looking out for.

One of the things that fascinates me about cryptozoological animals is their ability to show up in the most unlikely places, and this is certainly true of both Northumberland and Tyneside. Cryptids may, of course, hide away discreetly in the wilderness of the Northumbrian countryside, but they are equally as likely to be found waiting in a bus queue in Westgate Road, Newcastle. Trust me; I know about these things.

As places to launch a career in cryptozoology, Northumberland and Tyneside take some beating. The sheer diversity of the landscape – culturally, geologically, spiritually, socially and economically – provides an almost infinite number of niches in which cryptids – both conventional and truly paranormal – can thrive. I've said for many a year now that the north of England is an uncharted landscape when it comes to the Unknown. Few books, relatively speaking, have detailed with any success the bizarre occurrences which have taken place over the millennia – including the appearance of some exceedingly strange creatures that simply do not fit into the taxonomical tree that academics cling to with such passion. Personally, I wouldn't have it any other way. Of course, the industrialisation of huge swathes of land has reduced the likelihood that new creatures unknown to science may still live in the region, but the possibility hasn't been done away with altogether. From time to time, reports come in of unclassified butterflies and moths, hitherto unrecognised mice and even the odd (distinctly odd) feline. The gods may be cruel, but they can also be kind. They've gifted we Geordies with soul-inspiring mysteries that have yet to be investigated, let alone explained.

No one can even guess at the potential for cryptozoological research that exists in this region; the best we can say is that it is likely to be far greater than we can imagine. Intriguingly, there is a great dichotomy between the attitude of the general public to cryptozoological mysteries and that of the academic world. Academics in the north of England are, in the main, similar to their colleagues elsewhere; generally dismissive of the notion that weird beasties could be lolloping across the countryside and through our towns and villages without a permit. The public, however, are far more tolerant of the notion and do not seem in the least bit phased by the thought of winged dogs, ghost birds and giant lobsters. And why should they? Such creatures are part of our local heritage, and should be treasured.

This volume has merely scratched the surface of the north's cryptozoological history. Oh, and remember – if you do happen to see a giant rabbit, a gast adder or – heaven forbid – the shade of Wandering Willie down by Comical Corner, you read about it here first...

Bibliography

During my research for this volume I had occasion to consult over two hundred publications, some of them extremely rare and hard to locate. Not all the volumes listed below relate directly to the stories contained in this book; many are listed because they provide good background information concerning the north east of England and/or the culture of the peoples of the region. Others contain geographical, topological, biological and historical information that is also useful. Yet others give a good insight into the way Northerners view paranormal phenomena in general, and cryptozoology specifically. Some books listed below provide a global context to cryptozoological enigmas, and demonstrate a relationship between cryptids in the north east of England and other localities around the world.

Anderson, Roy; *The Violent Kingdom* (Butler Publishing [undated]).

Anon; *Chillingham Castle* (Chillingham Castle [undated]).

Anon; *Ghosts & Legends of Northumbria* (Sandhill Press, 1992).

Anon; *Myth & Magic of Northumbria* (Sandhill Press, 1992).

Anon; *Pictures of Cleadon Village – A Look Back in Time* (South Tyneside Libraries, 1994).

Anon; *The Castle of Newcastle upon Tyne* (Society of Antiquaries, Newcastle upon Tyne, 2000).

Armstrong, Pamela; *Dark Tales of Old Newcastle* (Bridge Studios, 1990).

Atkinson, Frank; *Life & Tradition in Northumberland & Durham* (Dalesman Books, 1986).

Banks, John; *Reminiscences of Smugglers & Smuggling* (Frank Graham, 1966).

Batey, Mavis; *The World of Alice* (Pitkin Guides Ltd, 1998).

Bath, Jo; *Dancing With the Devil* (Tyne Bridge Publishing, 2002).

Bidwell, P. T; *The Roman Fort of Arbeia at South Shields* (Tyne & Wear Museums, 1993).

Blackman, W. Haden; *North American Monsters* (Three Rivers Press, 1998).

Bord, Janet & Colin; *Alien Animals* (Book Club Associates, 1981).

Bottle, Erasmus; *A Boatload of History* (Stone Bottle Publications, 1999).

Brockie, William; *Legends & Superstitions of the County of Durham* (Sunderland, 1886).

Brookesmith, Peter [edit]; *Creatures From Elsewhere* (Macdonald & Co. Publishers, 1989).

Costello, Peter; *The Magic Zoo* (Sphere Books, 1979).

Downes, Jonathan & Freeman, Richard [edit]; *The CFZ Yearbook, 2004* (CFZ Press, 2007).

Downes, Jonathan; *Monster of the Mere* (CFZ Publications, 2002).

Downes, Jonathan; *The Owlman and Others* (CFZ Publications, 1997).

Edwards, Frank; *Stranger Than Science* (Lyle Stuart, 1959).

Farrant, David; *Beyond the Highgate Vampire* (British Psychic & Occult Society, 1997).

Freeman, Richard; *Dragons – More Than a Myth* (CFZ Press, 2005).

Freeman, Richard; *Exploring Dragons* (Explore Books, 2006).

Gerhard, Ken; *Big Bird!* (CFZ Press, 2007).

Hallowell, Michael J; *Ales & Spirits – The Haunted Pubs and Inns of South Tyneside* (The People's Press, 2001).

Henderson, William; *Folk Lore of the Northern Counties of England & the Borders* (EP Publishing, 1973).

Hill, Revd. Stuart G; *St. Peter's Church & the Wearmouth Jarrow Monastery* (St. Peter's DCC [undated]).

Kristen, Clive; *More Ghost Trails of Northumbria* (Casdec Ltd, 1993).

Legg, Gerald; T*he X-Ray Picture Book of Incredible Creatures* (Parrallel Books, 1995).

Mawnan-Peller, A: *Morgawr – The Monster of Falmouth Bay* (Morgawr Productions, 1976).

Michell, John & Rickard, Robert J M; *Living Wonders* (Thames & Hudson Ltd, 1982).

mult. au; *Creatures From Elsewhere* (Orbis Publishing Ltd, 1984).

mult. Au; *The Northumberland Village Book* (Countryside Books, Newbury and the NFWI, 1994).

Redfern, Nick; *Man-Monkey* (CFZ Press, 2007).

Ritson, Darren W; *Ghost Hunter; True Life Encounters From the North East* (GHP, 2006).

Ritson, Darren W; *In Search of Ghosts* (GHP, 2007).

Rogers, Ian; *Robert Ingham QC and His Friends* (South Tyneside Health Care Trust, 1995).

Sherman, Jory; *The Bamboo Demons* (NEL, 1979).

Shuker, Dr. Karl P. N; *Extraordinary Animals Revisited* (CFZ Press, 2007).

Woodhouse, Robert; *County Durham – Strange But True* (Sutton Publishing, 2004).

Wyndham, John; *The Kraken Wakes* (Michael Joseph Ltd, 1953).

Young, M. J; *A History of Catholic Jarrow* (Edward Thompson, 1921).

THE CENTRE FOR FORTEAN ZOOLOGY

So, what is the Centre for Fortean Zoology?

We are a non profit-making organisation founded in 1992 with the aim of being a clearing house for information, and coordinating research into mystery animals around the world. We also study out of place animals, rare and aberrant animal behaviour, and Zooform Phenomena; little-understood "things" that appear to be animals, but which are in fact nothing of the sort, and not even alive (at least in the way we understand the term).

Why should I join the Centre for Fortean Zoology?

Not only are we the biggest organisation of our type in the world, but - or so we like to think - we are the best. We are certainly the only truly global Cryptozoological research organisation, and we carry out our investigations using a strictly scientific set of guidelines. We are expanding all the time and looking to recruit new members to help us in our research into mysterious animals and strange creatures across the globe. Why should you join us? Because, if you are genuinely interested in trying to solve the last great mysteries of Mother Nature, there is nobody better than us with whom to do it.

What do I get if I join the Centre for Fortean Zoology?

For £12 a year, you get a four-issue subscription to our journal *Animals & Men*. Each issue contains 60 pages packed with news, articles, letters, research papers, field reports, and even a gossip column! The magazine is A5 in format with a full colour cover. You also have access to one of the world's largest collections of resource material dealing with cryptozoology and allied disciplines, and people from the CFZ membership regularly take part in fieldwork and expeditions around the world.

How is the Centre for Fortean Zoology organized?

The CFZ is managed by a three-man board of trustees, with a non-profit making trust registered with HM Government Stamp Office. The board of trustees is supported by a Permanent Directorate of full and part-time staff, and advised by a Consultancy Board of specialists - many of whom who are world-renowned experts in their particular field. We have regional representatives across the UK, the USA, and many other parts of the world, and are affiliated with other organisations whose aims and protocols mirror our own.

I am new to the subject, and although I am interested I have little practical knowledge. I don't want to feel out of my depth. What should I do?

Don't worry. We were *all* beginners once. You'll find that the people at the CFZ are friendly and approachable. We have a thriving forum on the website which is the hub of an ever-growing electronic community. You will soon find your feet. Many members of the CFZ Permanent Directorate started off as ordinary members, and now work full-time chasing monsters around the world.

I have an idea for a project which isn't on your website. What do I do?

Write to us, e-mail us, or telephone us. The list of future projects on the website is not exhaustive. If you have a good idea for an investigation, please tell us. We may well be able to help.

How do I go on an expedition?

We are always looking for volunteers to join us. If you see a project that interests you, do not hesitate to get in touch with us. Under certain circumstances we can help provide funding for your trip. If you look on the future projects section of the website, you can see some of the projects that we have pencilled in for the next few years.

In 2003 and 2004 we sent three-man expeditions to Sumatra looking for Orang-Pendek - a semi-legendary bipedal ape. The same three went to Mongolia in 2005. All three members started off merely subscribers to the CFZ magazine.

Next time it could be you!

Project Kerinci, Sumatra - 2003
In search of the bipedal ape Orang Pendek

How is the Centre for Fortean Zoology funded?

We have no magic sources of income. All our funds come from donations, membership fees, works that we do for TV, radio or magazines, and sales of our publications and merchandise. We are always looking for corporate sponsorship, and other sources of revenue. If you have any ideas for fund-raising please let us know. However, unlike other cryptozoological organisations in the past, we do not live in an intellectual ivory tower. We are not afraid to get our hands dirty, and furthermore we are not one of those organisations where the membership have to raise money so that a privileged few can go on expensive foreign trips. Our research teams both in the UK and abroad, consist of a mixture of experienced and inexperienced personnel. We are truly a community, and work on the premise that the benefits of CFZ membership are open to all.

What do you do with the data you gather from your investigations and expeditions?

Reports of our investigations are published on our website as soon as they are available. Preliminary reports are posted within days of the project finishing.

Each year we publish a 200 page yearbook containing research papers and expedition reports too long to be printed in the journal. We freely circulate our information to anybody who asks for it.

Is the CFZ community purely an electronic one?

No. Each year since 2000 we have held our annual convention - the *Weird Weekend* - in Exeter. It is three days of lectures, workshops, and excursions. But most importantly it is a chance for members of the CFZ to meet each other, and to talk with the members of the permanent directorate in a relaxed and informal setting and preferably with a pint of beer in one hand. Since 2006 - the *Weird Weekend* has been bigger and better and held in the idyllic rural location of Woolsery in North Devon. The 2008 event will be held over the weekend 15-17 August.

Since relocating to North Devon in 2005 we have become ever more closely involved with other community organisations, and we hope that this trend will continue. We also work closely with Police Forces across the UK as consultants for animal mutilation cases, and we intend to forge closer links with the coastguard and other community services. We want to work closely with those who regularly travel into the Bristol Channel, so that if the recent trend of exotic animal visitors to our coastal waters continues, we can be out there as soon as possible.

We are building a Visitor's Centre in rural North Devon. This will not be open to the general public, but will provide a museum, a library and an educational resource for our members (currently over 300) across the globe. We are also planning a youth organisation which will involve children and young people in our activities. We work closely with *Tropiquaria* - a small zoo in north Somerset, and have several exciting conservation projects planned.

Apart from having been the only Fortean Zoological organisation in the world to have consistently published material on all aspects of the subject for over a decade, we have achieved the following concrete results:

- Disproved the myth relating to the headless so-called sea-serpent carcass of Durgan beach in Cornwall 1975
- Disproved the story of the 1988 puma skull of Lustleigh Cleave
- Carried out the only in-depth research ever into the mythos of the Cornish Owlman
- Made the first records of a tropical species of lamprey
- Made the first records of a luminous cave gnat larva in Thailand.
- Discovered a possible new species of British mammal - the beech marten.
- In 1994-6 carried out the first archival fortean zoological survey of Hong Kong.
- In the year 2000, CFZ theories where confirmed when an entirely new species of lizard was found resident in Britain.
- Identified the monster of Martin Mere in Lancashire as a giant wels catfish
- Expanded the known range of Armitage's skink in the Gambia by 80%
- Obtained photographic evidence of the remains of Europe's largest known pike
- Carried out the first ever in-depth study of the *ninki-nanka*
- Carried out the first attempt to breed Puerto Rican cave snails in captivity
- Were the first European explorers to visit the `lost valley` in Sumatra
- Published the first ever evidence for a new tribe of pygmies in Guyana
- Published the first evidence for a new species of caiman in Guyana

EXPEDITIONS & INVESTIGATIONS TO DATE INCLUDE:

- 1998 Puerto Rico, Florida, Mexico *(Chupacabras)*
- 1999 Nevada *(Bigfoot)*
- 2000 Thailand *(Giant snakes called nagas)*
- 2002 Martin Mere *(Giant catfish)*
- 2002 Cleveland *(Wallaby mutilation)*
- 2003 Bolam Lake *(BHM Reports)*
- 2003 Sumatra *(Orang Pendek)*
- 2003 Texas *(Bigfoot; giant snapping turtles)*
- 2004 Sumatra *(Orang Pendek; cigau, a sabre-toothed cat)*
- 2004 Illinois *(Black panthers; cicada swarm)*
- 2004 Texas *(Mystery blue dog)*
- 2004 Puerto Rico *(Chupacabras; carnivorous cave snails)*
- 2005 Belize *(Affiliate expedition for hairy dwarfs)*
- 2005 Mongolia *(Allghoi Khorkhoi aka Mongolian death worm)*
- 2006 Gambia *(Gambo - Gambian sea monster , Ninki Nanka and Armitage s skink*
- 2006 Llangorse Lake *(Giant pike, giant eels)*
- 2006 Windermere *(Giant eels)*
- 2007 Coniston Water *(Giant eels)*
- 2007 Guyana *(Giant anaconda, didi, water tiger)*

To apply for a <u>FREE</u> information pack about the organisation and details of how to join, plus information on current and future projects, expeditions and events.

Send a stamped and addressed envelope to:

**THE CENTRE FOR FORTEAN ZOOLOGY
MYRTLE COTTAGE, WOOLSERY,
BIDEFORD, NORTH DEVON
EX39 5QR.**

or alternatively visit our website at:
www.cfz.org.uk

Other books available from
CFZ PRESS

THE OWLMAN AND OTHERS - 30th Anniversary Edition
Jonathan Downes - ISBN 978-1-905723-02-7

£14.99

EASTER 1976 - Two young girls playing in the churchyard of Mawnan Old Church in southern Cornwall were frightened by what they described as a "nasty bird-man". A series of sightings that has continued to the present day. These grotesque and frightening episodes have fascinated researchers for three decades now, and one man has spent years collecting all the available evidence into a book. To mark the 30th anniversary of these sightings, Jonathan Downes has published a special edition of his book.

DRAGONS - More than a myth?
Richard Freeman - ISBN 0-9512872-9-X

£14.99

First scientific look at dragons since 1884. It looks at dragon legends worldwide, and examines modern sightings of dragon-like creatures, as well as some of the more esoteric theories surrounding dragonkind.

Dragons are discussed from a folkloric, historical and cryptozoological perspective, and Richard Freeman concludes that: "When your parents told you that dragons don't exist - they lied!"

MONSTER HUNTER
Jonathan Downes - ISBN 0-9512872-7-3

£14.99

Jonathan Downes' long-awaited autobiography, *Monster Hunter*...

Written with refreshing candour, it is the extraordinary story of an extraordinary life, in which the author crosses paths with wizards, rock stars, terrorists, and a bewildering array of mythical and not so mythical monsters, and still just about manages to emerge with his sanity intact.......

MONSTER OF THE MERE
Jonathan Downes - ISBN 0-9512872-2-2

£12.50

It all starts on Valentine's Day 2002 when a Lancashire newspaper announces that "Something" has been attacking swans at a nature reserve in Lancashire. Eyewitnesses have reported that a giant unknown creature has been dragging fully grown swans beneath the water at Martin Mere. An intrepid team from the Exeter based Centre for Fortean Zoology, led by the author, make two trips – each of a week – to the lake and its surrounding marshlands. During their investigations they uncover a thrilling and complex web of historical fact and fancy, quasi Fortean occurrences, strange animals and even human sacrifice.

CFZ PRESS, MYRTLE COTTAGE,
WOOLFARDISWORTHY BIDEFORD,
NORTH DEVON, EX39 5QR
www.cfz.org.uk

Other books available from
CFZ PRESS

ONLY FOOLS AND GOATSUCKERS
Jonathan Downes - ISBN 0-9512872-3-0

£12.50

In January and February 1998 Jonathan Downes and Graham Inglis of the Centre for Fortean Zoology spent three and a half weeks in Puerto Rico, Mexico and Florida, accompanied by a film crew from UK Channel 4 TV. Their aim was to make a documentary about the terrifying chupacabra - a vampiric creature that exists somewhere in the grey area between folklore and reality. This remarkable book tells the gripping, sometimes scary, and often hilariously funny story of how the boys from the CFZ did their best to subvert the medium of contemporary TV documentary making and actually do their job.

WHILE THE CAT'S AWAY
Chris Moiser - ISBN: 0-9512872-1-4

£7.99

Over the past thirty years or so there have been numerous sightings of large exotic cats, including black leopards, pumas and lynx, in the South West of England. Former Rhodesian soldier Sam McCall moved to North Devon and became a farmer and pub owner when Rhodesia became Zimbabwe in 1980. Over the years despite many of his pub regulars having seen the "Beast of Exmoor" Sam wasn't at all sure that it existed. Then a series of happenings made him change his mind. Chris Moiser—a zoologist—is well known for his research into the mystery cats of the westcountry. This is his first novel.

CFZ EXPEDITION REPORT 2006 - GAMBIA
ISBN 1905723032

£12.50

In July 2006, The J.T.Downes memorial Gambia Expedition - a six-person team - Chris Moiser, Richard Freeman, Chris Clarke, Oll Lewis, Lisa Dowley and Suzi Marsh went to the Gambia, West Africa. They went in search of a dragon-like creature, known to the natives as `Ninki Nanka`, which has terrorized the tiny African state for generations, and has reportedly killed people as recently as the 1990s. They also went to dig up part of a beach where an amateur naturalist claims to have buried the carcass of a mysterious fifteen foot sea monster named 'Gambo', and they sought to find the Armitage's Skink (*Chalcides armitagei*) - a tiny lizard first described in 1922 and only rediscovered in 1989. Here, for the first time, is their story.... With an forward by Dr. Karl Shuker and introduction by Jonathan Downes.

BIG CATS IN BRITAIN YEARBOOK 2006
Edited by Mark Fraser - ISBN 978-1905723-01-0

£10.00

Big cats are said to roam the British Isles and Ireland even now as you are sitting and reading this. People from all walks of life encounter these mysterious felines on a daily basis in every nook and cranny of these two countries. Most are jet-black, some are white, some are brown, in fact big cats of every description and colour are seen by some unsuspecting person while on his or her daily business. 'Big Cats in Britain' are the largest and most active group in the British Isles and Ireland This is their first book. It contains a run-down of every known big cat sighting in the UK during 2005, together with essays by various luminaries of the British big cat research community which place the phenomenon into scientific, cultural, and historical perspective.

CFZ PRESS, MYRTLE COTTAGE,
WOOLSERY, BIDEFORD,
NORTH DEVON, EX39 5QR
w w w . c f z . o r g . u k

Other books available from
CFZ PRESS

THE SMALLER MYSTERY CARNIVORES OF THE WESTCOUNTRY
Jonathan Downes - ISBN 978-1-905723-05-8

£7.99

Although much has been written in recent years about the mystery big cats which have been reported stalking Westcountry moorlands, little has been written on the subject of the smaller British mystery carnivores. This unique book redresses the balance and examines the current status in the Westcountry of three species thought to be extinct: the Wildcat, the Pine Marten and the Polecat, finding that the truth is far more exciting than the currently held scientific dogma. This book also uncovers evidence suggesting that even more exotic species of small mammal may lurk hitherto unsuspected in the countryside of Devon, Cornwall, Somerset and Dorset.

THE BLACKDOWN MYSTERY
Jonathan Downes - ISBN 978-1-905723-00-3

£7.99

Intrepid members of the CFZ are up to the challenge, and manage to entangle themselves thoroughly in the bizarre trappings of this case. This is the soft underbelly of ufology, rife with unsavoury characters, plenty of drugs and booze." That sums it up quite well, we think. A new edition of the classic 1999 book by legendary fortean author Jonathan Downes. In this remarkable book, Jon weaves a complex tale of conspiracy, anti-conspiracy, quasi-conspiracy and downright lies surrounding an air-crash and alleged UFO incident in Somerset during 1996. However the story is much stranger than that. This excellent and amusing book lifts the lid off much of contemporary forteana and explains far more than it initially promises.

GRANFER'S BIBLE STORIES
John Downes - ISBN 0-9512872-8-1

£7.99

Bible stories in the Devonshire vernacular, each story being told by an old Devon Grandfather - 'Granfer'. These stories are now collected together in a remarkable book presenting selected parts of the Bible as one more-or-less continuous tale in short 'bite sized' stories intended for dipping into or even for bed-time reading. `Granfer` treats the biblical characters as if they were simple country folk living in the next village. Many of the stories are treated with a degree of bucolic humour and kindly irreverence, which not only gives the reader an opportunity to re-evaluate familiar tales in a new light, but do so in both an entertaining and a spiritually uplifting manner.

FRAGRANT HARBOURS DISTANT RIVERS
John Downes - ISBN 0-9512872-5-7

£12.50

Many excellent books have been written about Africa during the second half of the 19th Century, but this one is unique in that it presents the stories of a dozen different people, whose interlinked lives and achievements have as many nuances as any contemporary soap opera. It explains how the events in China and Hong Kong which surrounded the Opium Wars, intimately effected the events in Africa which take up the majority of this book. The author served in the Colonial Service in Nigeria and Hong Kong, during which he found himself following in the footsteps of one of the main characters in this book; Frederick Lugard – the architect of modern Nigeria.

**CFZ PRESS, MYRTLE COTTAGE,
WOOLFARDISWORTHY BIDEFORD,
NORTH DEVON, EX39 5QR
w w w . c f z . o r g . u k**

Other books available from
CFZ PRESS

ANIMALS & MEN - Issues 1 - 5 - In the Beginning
Edited by Jonathan Downes - ISBN 0-9512872-6-5

£12.50

At the beginning of the 21st Century monsters still roam the remote, and sometimes not so remote, corners of our planet. It is our job to search for them. The Centre for Fortean Zoology [CFZ] is the only professional, scientific and full-time organisation in the world dedicated to cryptozoology - the study of unknown animals. Since 1992 the CFZ has carried out an unparalleled programme of research and investigation all over the world. We have carried out expeditions to Sumatra (2003 and 2004), Mongolia (2005), Puerto Rico (1998 and 2004), Mexico (1998), Thailand (2000), Florida (1998), Nevada (1999 and 2003), Texas (2003 and 2004), and Illinois (2004). An introductory essay by Jonathan Downes, notes putting each issue into a historical perspective, and a history of the CFZ.

ANIMALS & MEN - Issues 6 - 10 - The Number of the Beast
Edited by Jonathan Downes - ISBN 978-1-905723-06-5

£12.50

At the beginning of the 21st Century monsters still roam the remote, and sometimes not so remote, corners of our planet. It is our job to search for them. The Centre for Fortean Zoology [CFZ] is the only professional, scientific and full-time organisation in the world dedicated to cryptozoology - the study of unknown animals. Since 1992 the CFZ has carried out an unparalleled programme of research and investigation all over the world. We have carried out expeditions to Sumatra (2003 and 2004), Mongolia (2005), Puerto Rico (1998 and 2004), Mexico (1998), Thailand (2000), Florida (1998), Nevada (1999 and 2003), Texas (2003 and 2004), and Illinois (2004). Preface by Mark North and an introductory essay by Jonathan Downes, notes putting each issue into a historical perspective, and a history of the CFZ.

BIG BIRD! Modern Sightings of Flying Monsters

£7.99

Ken Gerhard - ISBN 978-1-905723-08-9

From all over the dusty U.S./Mexican border come hair-raising stories of modern day encounters with winged monsters of immense size and terrifying appearance. Further field sightings of similar creatures are recorded from all around the globe. What lies behind these weird tales? Ken Gerhard is a native Texan, he lives in the homeland of the monster some call 'Big Bird'. Ken's scholarly work is the first of its kind. On the track of the monster, Ken uncovers cases of animal mutilations, attacks on humans and mounting evidence of a stunning zoological discovery ignored by mainstream science. Keep watching the skies!

STRENGTH THROUGH KOI
They saved Hitler's Koi and other stories

£7.99

Jonathan Downes - ISBN 978-1-905723-04-1

Strength through Koi is a book of short stories - some of them true, some of them less so - by noted cryptozoologist and raconteur Jonathan Downes. The stories are all about koi carp, and their interaction with bigfoot, UFOs, and Nazis. Even the late George Harrison makes an appearance. Very funny in parts, this book is highly recommended for anyone with even a passing interest in aquaculture, but should be taken definitely *cum grano salis*.

CFZ PRESS, MYRTLE COTTAGE,
WOOLSERY, BIDEFORD,
NORTH DEVON, EX39 5QR

Other books available from
CFZ PRESS

CFZ PRESS

BIG CATS IN BRITAIN YEARBOOK 2007
Edited by Mark Fraser - ISBN 978-1-905723-09-6

£12.50

People from all walks of life encounter mysterious felids on a daily basis, in every nook and cranny of the UK. Most are jet-black, some are white, some are brown; big cats of every description and colour are seen by some unsuspecting person while on his or her daily business. 'Big Cats in Britain' are the largest and most active research group in the British Isles and Ireland. This book contains a run-down of every known big cat sighting in the UK during 2006, together with essays by various luminaries of the British big cat research community.

CAT FLAPS! Northern Mystery Cats
Andy Roberts - ISBN 978-1-905723-11-9

£6.99

Of all Britain`s mystery beasts, the alien big cats are the most renowned. In recent years the notoriety of these uncatchable, out-of-place predators have eclipsed even the Loch Ness Monster. They slink from the shadows to terrorise a community, and then, as often as not, vanish like ghosts. But now film, photographs, livestock kills, and paw prints show that we can no longer deny the existence of these once-legendary beasts. Here then is a case-study, a true lost classic of Fortean research by one of the country's most respected researchers.

CENTRE FOR FORTEAN ZOOLOGY 2007 YEARBOOK
Edited by Jonathan Downes and Richard Freeman
ISBN 978-1-905723-14-0

£12.50

The Centre For Fortean Zoology Yearbook is a collection of papers and essays too long and detailed for publication in the CFZ Journal *Animals & Men*. With contributions from both well-known researchers, and relative newcomers to the field, the Yearbook provides a forum where new theories can be expounded, and work on little-known cryptids discussed.

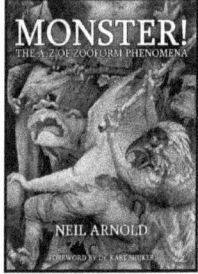

MONSTER! THE A-Z OF ZOOFORM PHENOMENA
Neil Arnold - ISBN 978-1-905723-10-2

£14.99

Zooform Phenomena are the most elusive, and least understood, mystery `animals`. Indeed, they are not animals at all, and are not even animate in the accepted terms of the word. Author and researcher Neil Arnold is to be commended for a groundbreaking piece of work, and has provided the world's first alphabetical listing of zooforms from around the world.

**CFZ PRESS, MYRTLE COTTAGE,
WOOLFARDISWORTHY BIDEFORD,
NORTH DEVON, EX39 5QR
www.cfz.org.uk**

Other books available from
CFZ PRESS

BIG CATS LOOSE IN BRITAIN
Marcus Matthews - ISBN 978-1-905723-12-6

£14.99

Big Cats: Loose in Britain, looks at the body of anecdotal evidence for such creatures: sightings, livestock kills, paw-prints and photographs, and seeks to determine underlying commonalities and threads of evidence. These two strands are repeatedly woven together into a highly readable, yet scientifically compelling, overview of the big cat phenomenon in Britain.

DARK DORSET
TALES OF MYSTERY, WONDER AND TERROR
Robert. J. Newland and Mark. J. North
ISBN 978-1-905723-15-6

£12.50

This extensively illustrated compendium has over 400 tales and references, making this book by far one of the best in its field. Dark Dorset has been thoroughly researched, and includes many new entries and up to date information never before published. The title of the book speaks for itself, and is indeed not for the faint hearted or those easily shocked.

MAN-MONKEY - IN SEARCH OF THE BRITISH BIGFOOT
Nick Redfern - ISBN 978-1-905723-16-4

£9.99

In her 1883 book, *Shropshire Folklore*, Charlotte S. Burne wrote: *'Just before he reached the canal bridge, a strange black creature with great white eyes sprang out of the plantation by the roadside and alighted on his horse's back'*. The creature duly became known as the `Man-Monkey`.

Between 1986 and early 2001, Nick Redfern delved deeply into the mystery of the strange creature of that dark stretch of canal. Now, published for the very first time, are Nick's original interview notes, his files and discoveries; as well as his theories pertaining to what lies at the heart of this diabolical legend.

EXTRAORDINARY ANIMALS REVISITED
Dr Karl Shuker - ISBN 978-1905723171

£14.99

This delightful book is the long-awaited, greatly-expanded new edition of one of Dr Karl Shuker's much-loved early volumes, *Extraordinary Animals Worldwide*. It is a fascinating celebration of what used to be called romantic natural history, examining a dazzling diversity of animal anomalies, creatures of cryptozoology, and all manner of other thought-provoking zoological revelations and continuing controversies down through the ages of wildlife discovery.

CFZ PRESS, MYRTLE COTTAGE,
WOOLFARDISWORTHY BIDEFORD,
NORTH DEVON, EX39 5QR
w w w . c f z . o r g . u k

Other books available from
CFZ PRESS

DARK DORSET CALENDAR CUSTOMS
Robert J Newland - ISBN 978-1-905723-18-8

£12.50

Much of the intrinsic charm of Dorset folklore is owed to the importance of folk customs. Today only a small amount of these curious and occasionally eccentric customs have survived, while those that still continue have, for many of us, lost their original significance. Why do we eat pancakes on Shrove Tuesday? Why do children dance around the maypole on May Day? Why do we carve pumpkin lanterns at Hallowe'en? All the answers are here! Robert has made an in-depth study of the Dorset country calendar identifying the major feast-days, holidays and celebrations when traditionally such folk customs are practiced.

CENTRE FOR FORTEAN ZOOLOGY 2004 YEARBOOK
Edited by Jonathan Downes and Richard Freeman
ISBN 978-1-905723-14-0

£12.50

The Centre For Fortean Zoology Yearbook is a collection of papers and essays too long and detailed for publication in the CFZ Journal *Animals & Men*. With contributions from both well-known researchers, and relative newcomers to the field, the Yearbook provides a forum where new theories can be expounded, and work on little-known cryptids discussed.

CENTRE FOR FORTEAN ZOOLOGY 2008 YEARBOOK
Edited by Jonathan Downes and Corinna Downes
ISBN 978 -1-905723-19-5

£12.50

The Centre For Fortean Zoology Yearbook is a collection of papers and essays too long and detailed for publication in the CFZ Journal *Animals & Men*. With contributions from both well-known researchers, and relative newcomers to the field, the Yearbook provides a forum where new theories can be expounded, and work on little-known cryptids discussed.

ETHNA'S JOURNAL
Corinna Newton Downes
ISBN 978 -1-905723-21-8

£9.99

Ethna's Journal tells the story of a few months in an alternate Dark Ages, seen through the eyes of Ethna, daughter of Lord Edric. She is an unsophisticated girl from the fortress town of Cragnuth, somewhere in the north of England, who reluctantly gets embroiled in a web of treachery, sorcery and bloody war...

**CFZ PRESS, MYRTLE COTTAGE,
WOOLFARDISWORTHY BIDEFORD,
NORTH DEVON, EX39 5QR
w w w . c f z . o r g . u k**

Other books available from
CFZ PRESS

ANIMALS & MEN - Issues 11 - 15 - The Call of the Wild
Jonathan Downes (Ed) - ISBN 978-1-905723-07-2

£12.50

Since 1994 we have been publishing the world's only dedicated cryptozoology magazine, *Animals & Men*. This volume contains fascimile reprints of issues 11 to 15 and includes articles covering out of place walruses, feathered dinosaurs, possible North American ground sloth survival, the theory of initial bipedalism, mystery whales, mitten crabs in Britain, Barbary lions, out of place animals in Germany, mystery pangolins, the barking beast of Bath, Yorkshire ABCs, Molly the singing oyster, singing mice, the dragons of Yorkshire, singing mice, the bigfoot murders, waspman, British beavers, the migo, Nessie, the weird warbling whatsit of the westcountry, the quagga project and much more...

IN THE WAKE OF BERNARD HEUVELMANS
Michael A Woodley - ISBN 978-1-905723-20-1

£9.99

Everyone is familiar with the nautical maps from the middle ages that were liberally festooned with images of exotic and monstrous animals, but the truth of the matter is that the *idea* of the sea monster is probably as old as humankind itself.

For two hundred years, scientists have been producing speculative classifications of sea serpents, attempting to place them within a zoological framework. This book looks at these successive classification models, and using a new formula produces a sea serpent classification for the 21st Century.

CENTRE FOR FORTEAN ZOOLOGY 1999 YEARBOOK
Edited by Jonathan Downes and Corinna Downes
ISBN 978 -1-905723-24-9

£12.50

The Centre For Fortean Zoology Yearbook is a collection of papers and essays too long and detailed for publication in the CFZ Journal *Animals & Men*. With contributions from both well-known researchers, and relative newcomers to the field, the Yearbook provides a forum where new theories can be expounded, and work on little-known cryptids discussed.

CENTRE FOR FORTEAN ZOOLOGY 1996 YEARBOOK
Edited by Jonathan Downes and Corinna Downes
ISBN 978 -1-905723-22-5

£12.50

The Centre For Fortean Zoology Yearbook is a collection of papers and essays too long and detailed for publication in the CFZ Journal *Animals & Men*. With contributions from both well-known researchers, and relative newcomers to the field, the Yearbook provides a forum where new theories can be expounded, and work on little-known cryptids discussed.

**CFZ PRESS, MYRTLE COTTAGE,
WOOLFARDISWORTHY BIDEFORD,
NORTH DEVON, EX39 5QR
w w w . c f z . o r g . u k**

Other books available from
CFZ PRESS

CFZ PRESS

BIG CATS IN BRITAIN YEARBOOK 2008
Edited by Mark Fraser - ISBN 978-1-905723-23-2

£12.50

People from all walks of life encounter mysterious felids on a daily basis, in every nook and cranny of the UK. Most are jet-black, some are white, some are brown; big cats of every description and colour are seen by some unsuspecting person while on his or her daily business. 'Big Cats in Britain' are the largest and most active research group in the British Isles and Ireland. This book contains a run-down of every known big cat sighting in the UK during 2007, together with essays by various luminaries of the British big cat research community.

CFZ EXPEDITION REPORT 2007 - GUYANA
ISBN 978-1-905723-25-6

£12.50

Since 1992, the CFZ has carried out an unparalleled programme of research and investigation all over the world. In November 2007, a five-person team - Richard Freeman, Chris Clarke, Paul Rose, Lisa Dowley and Jon Hare went to Guyana, South America. They went in search of giant anacondas, the bigfoot-like didi, and the terrifying water tiger.

Here, for the first time, is their story...With an introduction by Jonathan Downes and forward by Dr. Karl Shuker.

CENTRE FOR FORTEAN ZOOLOGY 2003 YEARBOOK
Edited by Jonathan Downes and Richard Freeman
ISBN 978-1-905723-19-5

£12.50

The Centre For Fortean Zoology Yearbook is a collection of papers and essays too long and detailed for publication in the CFZ Journal *Animals & Men*. With contributions from both well-known researchers, and relative newcomers to the field, the Yearbook provides a forum where new theories can be expounded, and work on little-known cryptids discussed.

CENTRE FOR FORTEAN ZOOLOGY 1998 YEARBOOK
Edited by Jonathan Downes and Graham Inglis
ISBN 978-1-905723-27-0

£12.50

The Centre For Fortean Zoology Yearbook is a collection of papers and essays too long and detailed for publication in the CFZ Journal *Animals & Men*. With contributions from both well-known researchers, and relative newcomers to the field, the Yearbook provides a forum where new theories can be expounded, and work on little-known cryptids discussed.

**CFZ PRESS, MYRTLE COTTAGE,
WOOLFARDISWORTHY BIDEFORD,
NORTH DEVON, EX39 5QR
w w w . c f z . o r g . u k**

Other books available from
CFZ PRESS

CENTRE FOR FORTEAN ZOOLOGY 2000-1 YEARBOOK
Edited by Jonathan Downes and Richard Freeman
ISBN 978-1-905723-28-7

£12.50

The Centre For Fortean Zoology Yearbook is a collection of papers and essays too long and detailed for publication in the CFZ Journal *Animals & Men*. With contributions from both well-known researchers, and relative newcomers to the field, the Yearbook provides a forum where new theories can be expounded, and work on little-known cryptids discussed.

CENTRE FOR FORTEAN ZOOLOGY 2002 YEARBOOK
Edited by Jonathan Downes and Richard Freeman
ISBN 978-1-905723-30-0

£12.50

The Centre For Fortean Zoology Yearbook is a collection of papers and essays too long and detailed for publication in the CFZ Journal *Animals & Men*. With contributions from both well-known researchers, and relative newcomers to the field, the Yearbook provides a forum where new theories can be expounded, and work on little-known cryptids discussed.

THE MYSTERY ANIMALS OF THE BRITISH ISLES: Northumberland and Tyneside
By Mike Hallowell ISBN 978-1-905723-29-4

£14.99

This is the first volume in a major new series from CFZ Press, which attempts nothing less than chronicling all the mystery animals, zooform phenomena, aberrations, and animal folklore of the United Kingdom and the Republic of Ireland. In this volume Mike introduces us to sea serpents, dragons, giant lobsters, ghost birds, the ghost of *Wandering Willie*, and even a vampire rabbit. A fantastic book and a great introduction to the series.

CENTRE FOR FORTEAN ZOOLOGY 1997 YEARBOOK
Edited by Jonathan Downes and Richard Freeman
ISBN 978-1-905723-31-7

The Centre For Fortean Zoology Yearbook is a collection of papers and essays too long and detailed for publication in the CFZ Journal *Animals & Men*. With contributions from both well-known researchers, and relative newcomers to the field, the Yearbook provides a forum where new theories can be expounded, and work on little-known cryptids discussed.

**CFZ PRESS, MYRTLE COTTAGE,
WOOLFARDISWORTHY BIDEFORD,
NORTH DEVON, EX39 5QR
www.cfz.org.uk**

www.ingramcontent.com/pod-product-compliance
Lightning Source LLC
Chambersburg PA
CBHW062156080426
42734CB00010B/1713